Butterflies th

MW00477322

# Butterflies
## OF THE CAROLINAS

### FIELD GUIDE

by Jaret C. Daniels

Adventure Publications, Inc.
Cambridge, MN

# ACKNOWLEDGEMENTS:

I would like to thank my wife, Stephanie, for her unending patience dealing with the countless number of caterpillars, butterflies and associated plant material that happens to always find a way into our home. Thanks to my parents for encouraging my early interest in biology.

Photo credits by photographer and page number:
**Cover photo: Black Swallowtail by Jaret C. Daniels**
**Tom Allen:** 44 (larva), 48 (larva), 50 (ventral, larva), 70 (larva), 86 (larva), 92 (larva), 102 (larva), 104 (larva), 108 (larva), 112 (larva), 116 (larva), 122 (larva), 126 (larva), 130 (larva), 132 (larva), 140 (larva), 142 (larva), 148 (larva), 160 (larva), 164 (larva), 168 (larva), 170 (larva), 174 (larva), 178 (larva), 182 (larva), 192 (larva), 194 (larva), 196 (larva), 198 (larva), 200 (larva), 210 (larva), 214 (larva), 230 (larva), 238 (larva), 250 (larva), 258 (larva), 274 (larva), 282 (larva), 294 (larva), 296 (larva), 298 (larva), 302 (larva), 306 (larva), 310 (larva), 312 (larva), 314 (larva), 316 (larva), 318 (larva), 324 (larva), 326 (larva), 328 (larva), 340 (ventral), 344 (larva), 358 (larva), 360 (larva), 384 (larva) **Rick & Nora Bowers/BowersPhoto.com:** 42, 44 (dorsal, ventral), 56 (both), 70 (male ventral), 82 (male), 84 (male), 100 (female), 110 (male dorsal), 120 (both), 142 (female), 154 (female dorsal), 166 (male dorsal), 204 (dorsal), 208 (dorsal), 220 (larva), 300 (male ventral), 286 (female), 290 (ventral, larva), 292 (male dorsal), 360 (female, ventral), 368 (male, female) **Randy Emmitt/www.RLEPhoto.com:** 48 (dorsal, ventral), 50 (dorsal), 90 (male, ventral), 92 (ventral), 116 (ventral), 118 (male, female), 122 (ventral), 124, 126 (ventral), 130 (male, female), 140 (dorsal), 146 (female dorsal, female ventral, male ventral), 148 (dorsal, ventral), 160 (dorsal, ventral), 164 (dorsal, ventral), 168 (dorsal, ventral), 176 (both), 182 (dorsal), 190 (female), 192 (male, female, ventral), 194 (female dorsal), 196 (male, female), 198 (ventral), 200 (dorsal, ventral, *loammi, loammi* ventral), 202, 210 (dorsal, ventral), 218 (all), 220 (dorsal), 230 (dorsal, ventral), 238 (ventral), 250 (ventral), 252 (ventral), 266 (ventral), 274 (ventral), 288 (male ventral), 290 (ventral), 294 (both), 296 (dorsal, ventral), 298 (male dorsal), 302 (ventral), 306 (ventral), 308 (male), 310 (male, female), 314 (ventral), 316 (male, female, ventral), 318 (male, female, ventral), 326 (male, ventral), 328 (ventral), 330 (dorsal, winter), 364 (male), 372, 374 (male, female) **Gary Meszaros/Dembinsky Photo Associates:** 334 (larva) **Jeffrey Glassberg:** 158 (both), 178 (dorsal, ventral), 174 (male dorsal), 198 (female), 258 (male ventral), 262 (both) **Paul Opler:** 174 (ventral) **John and Gloria Tveten:** 54 (all), 68 (larva), 84 (larva), 86 (ventral), 90 (larva), 102 (ventral), 104 (ventral), 108 (dorsal, ventral), 114 (green larva, red larva), 116 (dorsal), 128 (larvae), 136 (larva), 140 (ventral), 144, 162 (larva), 166 (ventral, larva), 170 (larva), 176 (ventral, larva), 188 (larva), 194 (ventral, "Pocahontas" ventral), 214 (dorsal, ventral), 216 (all), 228 (dorsal), 252 (ventral), 266 (larva), 270 (both), 272 (dorsal), 282 (dorsal), 286 (male ventral), 304 (male), 312 (ventral), 322 (male, larva), 324 (male, male ventral), 346 (larva), 348 (larva), 354 (ventral), 368 (ventral), 374 (larva), 394 (male ventral) **John and Gloria Tveten/KAC Productions:** 152, 212 **ATL, Elton Woodbury Collection:** 106 (both), 112 (ventral), 118 (larva), 136 (*ontario* ventral), 190 (larva), 322 (ventral), 340 (dorsal), 342 (larva), 374 (ventral) **Jaret C. Daniels:** all other photos

To the best of the publisher's knowledge, all photos except those of the Cofaqui Giant-Skipper were of live larvae and butterflies.

Book and Cover Design by Jonathan Norberg
Illustrations by Julie Martinez
Range Maps and Phenograms by Anthony Hertzel

# TABLE OF CONTENTS

**Introduction**

## WATCHING BUTTERFLIES
## IN THE CAROLINAS

People are rapidly discovering the joy of butterfly gardening and watching. Both are simple, fun and rewarding ways to explore the natural world and bring the beauty of nature closer. Few other forms of wildlife are more attractive than or as easy to observe as butterflies. Butterflies occur just about everywhere. They can be found from suburban gardens and urban parks to rural meadows and remote natural areas. Regardless of where you may live, there are a variety of butterflies to be seen. *Butterflies of the Carolinas Field Guide* is for those who wish to identify and learn more about the many different butterflies found in both North and South Carolina.

There are more than 725 species of butterflies in North America north of Mexico. While the majority of these are regular breeding residents, others show up from time to time as rare tropical strays. In the Carolinas, over 150 different butterflies have been recorded. Within that mix, there are widespread representatives that occur over a large portion of the continent and very rare species limited to only a few localized sites. Although such numbers pale in comparison to many tropical countries, the Carolinas boast a rich and diverse butterfly fauna. To aid in your exploration of this remarkable group of insects, this field guide covers all resident and stray species recorded within both states.

The breathtaking scenic beauty of the Carolinas makes them a premier vacation destination as well as a great place to observe and enjoy butterflies. But neither high mountain vistas nor sun-drenched beaches account for the abundant butterfly life. The Carolinas' rich fauna is directly related to elevation, climate and geography. Totaling over 83,000 square miles, the two states comprise a large portion of the eastern seaboard and contain three main physiographic regions. Bordering the west are the Appalachian Highlands that include the Great Smokey and Blue Ridge Mountains. At high elevations, this mountainous habitat represents the southernmost extension of the boreal forest characterized by evergreen stands of spruce and fir. At lower elevations, northern hardwood forests reminiscent of New England dominate. This region experiences four well-defined seasons. Here, summers are pleasantly cool with low humidity. Winters are often cold and snowy with extended periods of subfreezing temperatures. As a result, a variety of more typical northern species such as the Great Spangled Fritillary, Gray Comma and Silvery Blue can be found.

Further to the east is the Piedmont. This central region, also known as the Piedmont Plateau, is characterized by gently rolling foothills with an assortment of rivers, lakes, old fields, agricultural land and woodlands that comprise diverse natural communities. Here, four distinct seasons are still the rule. Shielded by the western mountains from cold interior air and tempered by the warm Atlantic to the east, year-round temperatures are generally more moderate.

Continuing eastward toward the sea is the Atlantic Coastal Plain. This primarily flat region extends along the many miles of coastline and includes a unique assortment of natural communities from wet pine savannas and bottomland hardwood forest to saltwater marshes and swamp forest. Blessed with abundant sunshine, relatively low elevation, and warmed by the nearby Gulf Stream waters of the Atlantic, the climate is considered subtropical. Here summers tend to be long, hot and humid. In contrast, winters are short and very mild with only brief periods of freezing temperatures. Combined with the Carolinas' geographic proximity to the Gulf States, a number of truly southern species are able to maintain year-round residency. Still other more tropical representatives such as the Zebra Longwing, Ceraunus Blue and Brazilian Skipper annually extend their ranges northward into the region to establish temporary breeding colonies.

For butterfly enthusiasts, the Carolinas are a natural amusement park. They offer both unmatched natural beauty and a diverse butterfly fauna. From the mountainous northwestern corner to the coastal southeastern tip, you can pass through over 100 distinct natural communities and encounter species typical of northern forests or occasional tropical vagrants. What more could you ask!

# WHAT ARE BUTTERFLIES?

Butterflies are insects. Along with moths, they comprise the Order Lepidoptera, a combination of Greek words meaning scale-winged, and can be differentiated from all other insects on that basis. Their four wings, as well as body, are typically almost entirely covered with numerous tiny scales. Overlapping like shingles on a roof, they make up the color and pattern of a butterfly's wings. Although generally wide and flat, some scales may be modified in shape, depending on the species and body location.

Butterflies and moths are closely related and often difficult to quickly tell apart. Nonetheless, there are some basic differences that are easy to identify even in the field. Generally, butterflies fly during the day, have large colorful wings that are held vertically together over the back when at rest, and bear distinctly clubbed antennae. In contrast, most moths are nocturnal. They are usually overall drabber in color and may often resemble dirty, hairy butterflies. At rest, they tend to hold their wings to the sides, and have feathery or threadlike antennae.

The following illustration points out the basic parts of a butterfly.

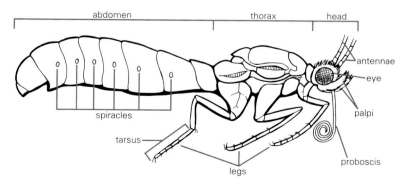

# BUTTERFLY BASICS

Adult butterflies share several common characteristics, including six jointed legs, two compound eyes, two antennae, a hard exoskeleton and three main body segments: the head, thorax and abdomen.

## Head

The head has two large compound eyes, two long clubbed antennae, a proboscis and two labial palpi. The rounded compound eyes are composed of hundreds of tiny individually lensed eyes. Together, they render a single, somewhat pixelated color image. Adult butterflies have good vision and are able to distinguish light in both the visible and ultraviolet range. Above the eyes are two long and slender antennae that are clubbed at the tip. They bear various sensory structures that help with orientation and smell. At the front of the head, below the eyes, are two protruding, hairy, brush-like structures called labial palpi. They serve to house and protect the proboscis, or tongue. The proboscis is a long flexible straw-like structure used for drinking fluids. It can be held tightly coiled below the head or extended when feeding. The length of a butterfly's proboscis determines the types of flowers and other foods from which it may feed.

## Thorax

Directly behind the head is the thorax. It is an enlarged muscular portion of the body divided into three segments that bear six legs and four wings. The front pair of legs may be significantly smaller and modified in certain families. Each leg is jointed and contains five separate sections, the last of which is the tarsus (pl. tarsi) or foot, which bears a tiny, hooked claw at the end. In addition to enabling the butterfly to securely grasp leaves, branches or other objects, the tarsi have sensory structures that are used to taste. Adult females scratch a leaf surface with their front tarsi to release the leaf's chemicals and taste whether they have found the correct host plant. Above the legs are two pairs of wings. Made up of two thin membranes supported by rigid veins, the generally large, colorful wings are covered with millions of tiny scales that overlap like shingles on a roof. The wings are used for a variety of critical functions, including flight, thermoregulation, sex recognition, camouflage, mimicry and predator deflection.

## Abdomen

The last section of a butterfly's body is the abdomen. This long, slender portion is comprised of ten segments and contains the reproductive, digestive and excretory systems along with a series of small lateral holes, called spiracles, for air exchange. The reproductive organs or genitalia are located at the end of the abdomen. Male butterflies have two modified structures called claspers that are used to grasp the female during copulation. Females possess a genital opening for mating and a second opening for egg laying. While these structures are often difficult to see in certain species, females generally have a much larger, "fatter-looking" abdomen because they carry a large complement of eggs.

## THE BUTTERFLY LIFE CYCLE

All butterflies pass through a life cycle consisting of four developmental stages: egg, larva, pupa and adult. Regardless of adult size, they begin life as a small egg. A female butterfly may lay her eggs singly, in small clusters or in large groupings on or near the appropriate host plant. Once an egg hatches, the tiny larva begins feeding almost immediately. Butterfly larvae are herbivores—with the exception of Harvester larvae, which eat aphids—and they essentially live to eat. As a result, they can grow at an astonishing rate. Unlike humans, though, all insects including butterflies have an external skeleton. In order to grow, a developing caterpillar (or larva) must shed its skin, or molt, several times during its life. Each time the larva does, it discards its old, tight skin to make room for the new, roomier and often different-looking skin underneath. These different stages in a caterpillar's growth are called instars. Once fully grown, the larva stops eating and seeks a safe place to pupate. It usually attaches itself to a branch, twig or other surface with silk and molts for the last time to reveal the pupa (or chrysalis). Inside, the larval structures are broken down and reorganized into the form of an adult butterfly. At the appropriate time, the pupa splits open and a beautiful new butterfly emerges. The adult hangs quietly and begins to expand its crumpled wings by slowly forcing blood through the veins. After a few hours, its wings are fully hardened and the butterfly is ready to fly.

## BUTTERFLY FAMILIES

Butterflies can be divided into five major families: Hesperiidae, Lycaenidae, Nymphalidae, Papilionidae and Pieridae. The members of each family have certain basic characteristics and behaviors that can be particularly useful for identification. Keep in mind that the features listed are only generalities, not hard and fast rules, and that there may be individual exceptions.

### Hesperiidae: Skippers

***Skippers*** are small- to medium-sized butterflies with robust, hairy bodies and relatively compact wings. They are generally brown, orange or white and their antennae bear short, distinct hooks at the tip. Adults have a quick and erratic flight, usually low to the ground. Skippers are divided into three main subfamilies: banded skippers (Subfamily Hesperiinae), giant-skippers (Subfamily Megathyminae) and spread-wing skippers (Subfamily Pyrginae).

***Banded skippers*** (Subfamily Hesperiinae) are small brown or orange butterflies with somewhat pointed forewings. Many have dark markings or distinct black forewing stigmas. They readily visit flowers and hold their wings together over the back while feeding. Adults often perch or rest in a characteristic posture with forewings held partially open and hindwings separated and lowered further.

As their name suggests, ***giant-skippers*** (Subfamily Megathyminae) dwarf most other members of the family. They are medium-sized brown butterflies with yellow markings and thick, robust bodies. The adults have a fast and rapid flight. Males establish territories and generally perch on low vegetation. Adults do not visit flowers.

***Spread-wing skippers*** (Subfamily Pyrginae) are generally dark, dull-colored butterflies with wide wings. Most have small, light spots on the forewings. Some species have hindwing tails. The adults often feed, rest and perch with their wings outstretched. They readily visit flowers.

## Lycaenidae: Gossamer Wings

This diverse family includes harvesters, blues, hairstreaks and metalmarks. The adults are small and often brilliantly colored but easily overlooked. Within the family is the Carolinas' smallest butterfly, the Eastern Pygmy-Blue. Throughout the state, blues and hairstreaks predominate, with harvesters, metalmarks and coppers represented by only one species each.

*Coppers* (Subfamily Lycaeninae) are small, sexually dimorphic butterflies found primarily in temperate regions. As their name suggests, the upper wing surfaces of most species are ornately colored with metallic reddish orange or purple. Many eastern species are associated with moist habitats including bogs, wet meadows and marshes. Populations are often quite localized but may be numerous when encountered. Adults typically scurry close to the ground with a quick flight and frequently visit available flowers or perch on low-growing vegetation with their wings held partially open.

The *harvester* (Subfamily Miletinae) is the only North American member of this unique, primarily Old World, subfamily. It holds the distinction as our only butterfly with carnivorous larvae. Instead of feeding on plant leaves, the larvae devour woolly aphids. The adults do not visit flowers, but sip honeydew, a sugary secretion produced by their host aphids.

Aptly named, *blues* (Subfamily Polyommatinae) are generally bright blue on the wings above. The sexes differ and females may be brown or dark gray. The wings beneath are typically whitish gray with dark markings and distinct hindwing eyespots. The eyes are wrapped around the base of the antennae. The palpi are reduced and close to the head. Adults have a moderately quick and erratic flight, usually low to the ground. At rest, they hold their wings together over the back. Males frequently puddle at damp ground. Most blues are fond of open, disturbed sites with weedy vegetation.

**Metalmarks** (Subfamily Riodininae) are characterized by metallic flecks of color or even overall metallic-looking wings. They have eyes entirely separate from the antennal bases, and the palpi are quite prominent. Metalmarks reach tremendous diversity in the tropics where they come in an almost unending array of colors and patterns. By contrast, most U.S. species are small rust, gray or brownish butterflies. They characteristically perch with their colorful wings outstretched and may often land on the underside of leaves, especially when disturbed. Adults have a low, scurrying flight. Several species are of conservation concern.

**Hairstreaks** (Subfamily Theclinae) tend to be larger in size than blues. The wings below are often intricately patterned and bear colorful eyespots adjacent to one or two small, distinct hair-like tails on each hindwing. The adults have a quick, erratic flight and can be a challenge to follow. They regularly visit flowers and hold their wings together over the back while feeding and at rest. Additionally, they have a unique behavior of moving their hindwings up and down when perched. The sexes regularly differ. Hairstreaks can be found in a wide range of habitats from dense woodlands to open, disturbed sites. Many species have a single spring generation.

## Nymphalidae: Brush-Foots

**Brush-foots** are the largest and most diverse family of butterflies. Its members include emperors, leafwing butterflies, milkweed butterflies, longwing butterflies, snouts, true brush-foots and satyrs and wood nymphs. In all, the first pair of legs is significantly reduced and modified into small brush-like structures, giving the family its name and making it look as if they only have four legs.

**Emperors** (Subfamily Apaturinae) are medium-sized and brownish with short, stubby bodies and a robust thorax. Their wings typically have dark markings and small dark eyespots. The adults are strong and rapid fliers. Males establish territories and perch on tree trunks or overhanging branches. At rest, they hold their wings together over the back. They feed on dung, carrion, rotting fruit or tree sap and do not visit flowers. Emperors inhabit rich woodlands and rarely venture far into open areas. They are nervous butterflies and difficult to closely approach.

**Leafwings** (Subfamily Charaxinae) are medium-sized butterflies with irregular wing margins. They are bright tawny orange above, but mottled gray to brown below and resemble a dead leaf when resting with their wings closed. The adults have a strong, rapid and erratic flight. They are nervous butterflies and difficult to closely approach. Males establish territories and perch on tree trunks or overhanging branches. Individuals also often land on the ground. They feed on dung, carrion, rotting fruit or tree sap and do not visit flowers.

**Milkweed butterflies** (Subfamily Danainae) are large butterflies with boldly marked black and orange wings. Their flight is strong and swift with periods of gliding. Adults are strongly attracted to flowers and feed with their wings folded tightly over the back. Males have noticeable black scent patches in the middle of each hindwing.

**Longwing butterflies** (Subfamily Heliconiinae) are colorful, medium- to large-sized butterflies with narrow, elongated wings, slender bodies, long antennae and large eyes. The flight varies from quick and erratic to slow and fluttering. The adults readily visit flowers and nectar with their wings outstretched. At rest, they hold their wings together over the back. Most individuals are long-lived.

**Snouts** (Subfamily Libytheinae) are medium-sized, generally drab brown butterflies representing about ten species worldwide with only one found in the U.S. They have extremely elongated labial palpi, reduced front legs and cyptically colored ventral hindwings. They rest with their wings tightly closed and resemble dead leaves. The American Snout often undergoes massive migrations.

**True brush-foots** (Subfamily Nymphalinae) are colorful, small- to medium-sized butterflies with no overall common wing shape. Most have stubby, compact bodies and a robust thorax. The adults have a strong, quick flight, usually low to the ground. Most are nervous and often difficult to approach. At rest, they hold their wings together over the back. Many are attracted to flowers while others feed on dung, carrion, rotting fruit or tree sap.

*Satyrs* and ***wood nymphs*** (Subfamily Satyrinae) are small- to medium-sized drab brown butterflies. Their wings are marked with dark stripes and prominent eyespots. The adults have a slow, somewhat bobbing flight usually low to the ground. They inhabit shady woodlands and adjacent open, grassy areas. The adults rarely visit flowers. They are instead attracted to dung, carrion, rotting fruit or tree sap. At rest, they hold their wings together over the back and are generally easy to closely approach. They regularly land on the ground.

## Papilionidae: Swallowtails

***Swallowtails*** are easily recognized by their large size and noticeably long hindwing tails. Within the Carolinas, only the Polydamas Swallowtail lacks tails. They are generally dark in color with bold markings. The adults have a swift and powerful flight, usually several meters off the ground, and regularly visit flowers. Most swallowtails continuously flutter their wings while feeding. Males often puddle at damp ground. They generally are found in and along woodland areas and adjacent open sites.

## Pieridae: Sulphurs and Whites

Members of the family are small- to medium-sized butterflies. As their name suggests, the adults are typically some shade of white or yellow. Many have dark markings. Most species are sexually dimorphic and seasonally variable. The adults have a moderately quick and erratic flight, usually low to the ground. They are fond of flowers and hold their wings together over the back while feeding. Males often puddle at damp ground. ***Sulphurs*** (Subfamily Coliadinae) and ***whites*** (Subfamily Pierinae) are common butterflies of open, disturbed sites where their weedy larval host plants abound.

# OBSERVING BUTTERFLIES IN THE FIELD

While butterflies are entertaining and beautiful to watch, correctly identifying them can often be a challenge. But it's generally not as difficult as it might seem. With a little practice and some basic guidelines, you can quickly learn to peg that unknown butterfly.

One of the first and most obvious things to note when you spot a butterfly is its size. You will discover that butterflies generally come in one of three basic dimensions: small, medium and large. This system may sound ridiculously arbitrary at first, but when you begin to regularly observe several different butterflies together in the field, these categories quickly start to make sense. For a starting point, follow this simple strategy: the next time you see a Monarch butterfly, pay close attention to its size. You may wish to use your hand as a reference. Most Monarchs have a wingspan close to the length of your palm (about four inches) as measured from the base of your fingers to the start of your wrist. This is considered a large butterfly. From here it's basically a matter of fractions. A medium-sized butterfly by comparison would have a wingspan of generally about half that size (about two inches). Finally, a butterfly would be considered small if it had a wingspan one quarter that of a Monarch (around one inch).

Next, pay close attention to color and pattern of the wings. This field guide is organized by color, and you can quickly navigate to the appropriate section. First, start by noting the overall ground color. Is it, for example, primarily black, yellow, orange or white? Then try to identify any major pattern elements. Does the butterfly have any distinct stripes, bands or spots? Depending on the behavior of the butterfly in the wild, keep in mind that the most visible portion of a butterfly may either be the upper surface of the wings (dorsal surface) or the underside of the wings (ventral surface). If you have a particularly cooperative subject, you may be able to closely observe both sides. Finally, carefully note the color and position of any major markings. For example, if the butterfly has a wide yellow band on the forewing, is it positioned in the middle of the wing or along the outer edge? Lepidopterists have a detailed vocabulary for wing pattern positions. The following illustrations of general wing features and wing areas should help you become familiar with some terminology.

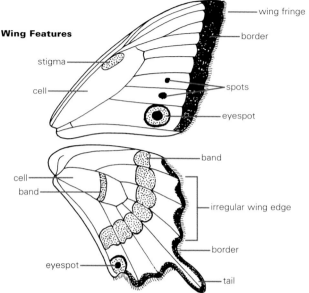

**Wing Features**

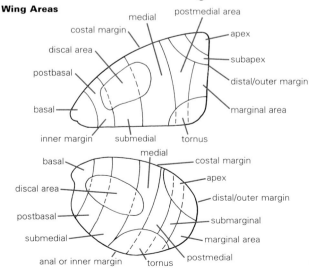

**Wing Areas**

Next, note the shape of the butterfly's wings, particularly the forewings. Are they generally long and narrow, rounded, broad, pointed or angled? Butterflies such as the Gulf Fritillary and Zebra Longwing have noticeably elongated wings. Others, like the Carolina Satyr and Eastern Tailed-Blue, have short, generally rounded wings. Next, do the wings have any unique features? Many swallowtails and hairstreaks, for instance, have distinct hindwing tails, while Question Marks and Eastern Commas have visibly irregular wing margins. Clues like this can help you quickly narrow the butterfly down to a particular family or distinguish it from a similarly colored species.

The way a butterfly flies may also be useful for identification. While it is generally difficult to easily pick up particular features or color patterns when a butterfly is moving, its flight pattern can often be very distinctive. Carefully follow the butterfly as it flies and watch how it behaves in the air. Is it soaring above your head or scurrying rapidly along the ground? Is it moving fast and erratically, or fluttering slowly about? Monarchs, for instance, have a very unique flight pattern. They flap their wings quickly several times, glide for bit, and then quickly flap their wings again. Other butterflies, such as most wood nymphs and satyrs, have a characteristic low, bobbing flight.

Sometimes you can gain important clues about a butterfly by the way it behaves when feeding. Next time you see a butterfly feeding, watch its wings. Does it hold them tightly closed, spread them wide open, or move them in a fluttering motion? Most swallowtails, for example, continuously flutter their wings. This behavior is a quick and reliable diagnostic that can be seen from a fair distance.

Note the habitat in which the butterfly occurs. Is it darting between branches along a shady and moist woodland path? Perched on the top of a grass blade in a saltwater marsh? Bobbing among low grasses in a wet prairie? Fluttering from one flower to the next in a fallow agricultural field? Many butterflies have strong habitat preferences. Some are restricted to a single particular habitat while others may occur in a wide range of habitats.

Sometimes even the date can offer a useful hint. Many hairstreaks, for instance, are univoltine, meaning they produce just one generation. As a result, the adults occur only during a narrow window of time each spring. Similarly, several butterflies overwinter as adults. They may be active at times when few other species are around.

Butterfly observation and identification are skills, and it takes time and practice to master them. To speed up the learning curve, you may also wish to join a local butterfly gardening or watching club or society. The members can help give advice, accompany you in the field, or share directions to great butterfly watching spots.

## DETERMINING A BUTTERFLY'S COLOR

To help make butterfly field identification fun and easy, this guide is organized by color, allowing you to quickly navigate to the appropriate section. In addition, smaller butterflies are always toward the beginning of each section, and the largest are toward the back. Butterflies are not static subjects, so determining their color can be a bit tricky. Several factors such as age, seasonal variation and complex wing and color patterns may affect your perception of a butterfly's color.

### Age

Adult age can influence a butterfly's color. While butterflies don't get wrinkles or gray hairs, they do continue to lose wing scales during their life. This type of normal wing wear combined with more significant wing damage can cause once vibrant colors or iridescence to fade and pattern elements to become less distinct. As a result, the bright tawny orange wings of a freshly emerged Great Spangled Fritillary may appear dull orange or almost yellowish in an old, worn individual.

### Seasonal Variation

Time of year can play a role as well. Some butterflies produce distinct seasonal forms that may vary significantly in color and to a lesser extent in wing shape or overall size. For the Sleepy Orange, the change is dramatic. Individuals produced during the warm summer months are bright butter yellow on the wings below. Winter-forms, by contrast, have dark rust-colored ventral hindwings perfect for blending into a predominantly brown fall and winter landscape. For species that display extensive seasonal variation, the differences are discussed and every attempt is made to include an image of both forms. Since you'll probably be looking for butterflies in the warmer months, pictures of winter-forms are insets and not main images.

## Complex Wing and Color Patterns

For many butterflies, wing color and pattern can vary tremendously between the dorsal and ventral surface. The wings of a Great Purple Hairstreak for example are brilliant iridescent blue above and dull black below. The bright dorsal coloration is readily visible during flight but concealed at rest when the hairstreak holds its wings firmly closed. As a result, the butterfly may appear either blue or black depending on your perspective and the butterfly's activity.

Most hairstreaks (subfamily Theclinae) and sulphurs (subfamily Coliadinae) typically feed and rest with their wings closed. As a result, very few photographs of the dorsal wing surface of free-flying butterflies in these groups exist. Similarly, most blues (subfamily Polyommatinae) perch and feed with their wings closed or hold them only partially open when basking. Many of these same butterflies are also two-toned (blue on the dorsal surface, but whitish to grubby gray below; or orange on the dorsal surface, but much yellower below). But because a large number of these butterflies are so small, chances are that you'll notice them first when they fly and reveal their brighter dorsal coloration. Two- toned butterflies such as these have been placed in the color section that reflects the brighter coloration that you're most likely to notice first.

For example, if you're looking in the white section for a small butterfly with lots of bands and spots on the ventral surface, but can't find it, try the blue section. You might be looking at a perched Spring Azure, which is bright blue above but a rather nondescript brownish gray white below.

Color can vary slightly even among butterflies of the same

species. If you see a butterfly that appears to be tawny colored, but you can't find it in the orange section, try the brown section. You might be looking at a Tawny Emperor, which can be perceived as orange or brown, depending on the individual.

Finally, females are generally more drab than males of the same species. They can have less iridescent color, fewer or darker markings, and can appear overall darker or paler. For some species, such as the Diana Fritillary and Eastern Tailed-Blue, the females and males look very different and are pictured in separate color sections.

## COLOR SECTION TROUBLESHOOTING

With all the variables that can impact the perception of a butterfly's color, identifying these creatures may seem daunting. Taking the time to check a color section or two is worth being able to positively identify a mystery butterfly.

**Can't find it in...          Try looking in...**

### Black                      Brown

In certain conditions, some of the dark brown butterflies, especially swallowtails and skippers, might appear black

### Blue                       Blue or Black

Blues (subfamily Polyommatinae) often rest or feed with their wings closed, so the blue dorsal coloration is visible primarily during flight or when the individual is basking. Check the illustration and description for field marks. Some black butterflies have varying amounts of iridescent blue scales; they may look blue in certain conditions.

### Brown                      Orange

Many butterflies described as tawny can be perceived as either brown or orange.

### Gray                       Brown

Many of the Hairstreaks (subfamily Theclinae) have light colored ventral wings that can vary in the amount of gray or brown they show.

### Green                      White

Some of the female Sulphurs (subfamily Coliadinae) have a common white form that may appear to be greenish.

### Orange                    Brown

Many of the skippers are brown but have bright orange field marks. Other butterflies are tawny colored, so could be perceived as either brown or orange.

### White                     Blue

Blues (subfamily Polyommatinae) perch with their wings closed, so appear to be white butterflies.

### Yellow                    Orange

Some of the Sulphurs (subfamily Coliadinae) have bright orange dorsal wings.

## BUTTERFLY GARDENING

One of the easiest ways to observe local butterflies is to plant a butterfly garden. Even a small area can attract a great variety of species directly to your yard. For best results, include both adult nectar sources and larval host plants. Most adult butterflies are generalists and will visit a broad range of colorful flowers in search of nectar. Developing larvae, on the other hand, typically have very discriminating tastes and often rely on only a few very specific plant species for food. For assistance with starting a butterfly garden, seek the guidance of local nursery professionals. They can help you determine which plants will grow well in your area. This field guide provides a list of larval host plants for each species, as well as a listing of good nectar plants. These lists begin on page 402.

# BUTTERFLY Q & A:
## *What's the difference between a butterfly and a moth?*

While there is no one simple answer to this question, butterflies and moths generally differ based on their overall habits and structure. Butterflies are typically active during the day (diurnal), while moths predominantly fly at night (nocturnal). Butterflies possess slender antennae that are clubbed at the end. Those of moths vary from long narrow filaments to broad fern-like structures. At rest, butterflies tend to hold their wings together vertically over the back. Moths rest with their wings extended flat out to the sides or folded alongside the body. Butterflies generally have long, smooth and slender bodies. The bodies of moths are often robust and hairy. Finally, most butterflies are typically brightly colored, while most moths tend to be dark and somewhat drab.

## *What can butterflies see?*

Butterflies are believed to have very good vision and to see a single color image. Compared to humans, they have an expanded range of sensitivity and are able to distinguish wavelengths of light into the ultraviolet range.

## *Do butterflies look the same year-round?*

Many butterflies produce distinct seasonal forms that differ markedly in color, size, reproductive activity and behavior. Good examples within the Carolinas include the Common Buckeye, Barred Sulphur and Sleepy Orange. The seasonal forms are determined based on a variety of environmental cues (temperature, rainfall, day length) that immature stages experience during development. Warm summer temperatures and long days forecast conditions that are highly favorable for continued development and reproduction. Summer-form individuals are generally lighter in color, short-lived, and reproductively active. As fall approaches, cooler temperatures and shortening day lengths mean future conditions may be unfavorable for continued development and reproduction. Winter-form adults are generally darker, display increased pattern elements, larger in size, longer lived, and survive the winter months in a state of reproductive diapause.

### Do caterpillars have eyes?

Yes, caterpillars or larvae generally have six pairs of simple eyes called ocelli. They are able to distinguish basic changes in light intensity but are believed to be incapable of forming an image.

### How do caterpillars defend themselves?

Caterpillars or larvae are generally plump, slow moving creatures that represent an inviting meal for many predators. To protect themselves, caterpillars employ a variety of different strategies. Many, like the Zebra Longwing or Pipevine Swallowtail, sequester specific chemicals from their host plants that render them highly distasteful or toxic. These caterpillars are generally brightly colored to advertise their unpalatability. Others rely on deception or camouflage to avoid being eaten. Palamedes Swallowtail larvae have large, conspicuous false eyes on an enlarged thorax that help them resemble a small snake or lizard. By contrast, larvae of the Georgia Satyr are solid green and extremely well camouflaged against the green leaves of their host. Some larvae conceal their whereabouts by constructing shelters. American Painted Lady larvae weave leaves and flowerheads together with silk and rest safely inside when not actively feeding. Still others have formidable spines and hairs or produce irritating or foul-smelling chemicals to deter persistent predators.

### Do butterfly caterpillars make silk?

Yes, butterfly larvae produce silk. While they don't typically spin an elaborate cocoon around their pupa like moths, they use silk for a variety of purposes, including the construction of shelters, anchoring or attaching their chrysalids, and to gain secure footing on leaves and branches.

### What happens when a butterfly's scales rub off?

Contrary to the old wives' tale, if you touch a butterfly's wing and remove scales in the process it is still capable of flying. In fact, a butterfly typically continuously loses scales during its life from normal wing wear. Scales serve a variety of purposes, from thermoregulation and camouflage to pheromone dispersal and species or sex recognition, but are not critical for flight. Once gone, however, the scales are permanently lost and will not grow back.

## Why do butterflies gather at mud puddles?

Adult butterflies are often attracted to damp or moist ground and may congregate at such areas in large numbers. In most cases, these groupings, or "puddle clubs," are made up entirely of males. They drink from the moisture to gain water and salts (sodium ions) that happen to come into solution. This behavior helps males replenish the sodium ions lost when they pass a packet of sperm and accessory gland secretions to the female during copulation. The transferred nutrients have been shown to play a significant role in egg production and, thus, female reproductive output.

## How long do butterflies live?

In general, most butterflies are extremely short-lived and survive in the wild for an average of about two weeks. There are, of course, numerous exceptions to this rule. The Zebra Longwing is a perfect example. Adults of this butterfly may survive for 4–6 months. Still others, particularly species that migrate long distances and/or overwinter as adults, are capable of surviving for extended periods of time.

## Where do butterflies go at night?
## Where do they go during a rainstorm?

In the evening or during periods of inclement weather, most butterflies seek shelter under the leaves of growing plants or among vegetation.

## Do all butterflies visit flowers?

All butterflies are fluid feeders. While a large percentage of them rely on sugar-rich nectar as the primary energy source for flight, reproduction and general maintenance, many species also feed on, or exclusively utilize, the liquids and dissolved nutrients produced by other food resources such as dung, carrion, rotting fruit or vegetation, sap and bird droppings.

## Do butterflies grow?

No. An adult butterfly is fully grown upon emergence from its chrysalis.

## How do caterpillars grow?

A caterpillar's job in life is to eat and grow. But larvae, like all other insects, have an external skeleton. Therefore, in order to increase in size, they must shed their skin, or molt, several times during development. Essentially, their skin is like a trash bag. It is packed full of food until there is no more room. Once full, it is discarded for a larger, baggier one underneath and the process continues.

## What eats butterflies?

Butterflies face an uphill battle for survival. Out of every one hundred eggs produced by a female butterfly, approximately only one percent survive to become an adult. And as an adult, the odds don't get much better. Various birds, small mammals, lizards, frogs, toads, spiders and other insects all prey on butterflies.

## What's the difference between a chrysalis and a cocoon?

Once fully grown, both moth and butterfly caterpillars molt a final time to form a pupa. Most moths surround their pupae with a constructed silken case called a cocoon. By contrast, a butterfly pupa, frequently termed a chrysalis, is generally naked. In most cases, butterfly chrysalids are attached to a leaf, twig or other surface with silk. In some instances, they may be unattached or surrounded by a loose silken cocoon.

## Are all butterfly scales the same?

No. The scales on a butterfly's wings and body come in a variety of different sizes and shapes. Some may be extremely elongated and resemble hairs while others are highly modified for the release of pheromones during courtship. Those responsible for making up wing color and pattern are generally wide and flat. They are attached at the base and overlap like shingles on a roof. The colors we see are the result of either pigments contained in the scales or the diffraction of light caused by scale structure. Iridescent colors such as blue, green, purple and silver usually result from scale structure. While pigmented scales are the norm, many species have a combination of both types on their wings.

# BUTTERFLY QUICK-COMPARE

Common Sootywing
pg. 43

Grizzled Skipper
pg. 45

Great Purple Hairstreak
pg. 47

Texan Crescent
pg. 49
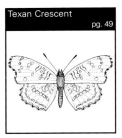

Baltimore Checkerspot
pg. 51

Red Admiral
pg. 53
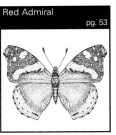

Milbert's Tortoiseshell
pg. 55
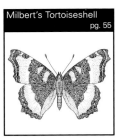

Mangrove Skipper
pg. 57

Zebra Longwing
pg. 59
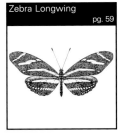

Red-spotted Purple
pg. 61

Zebra Swallowtail
pg. 63

Black Swallowtail
pg. 65

### Pipevine Swallowtail
pg. 67

### Mourning Cloak
pg. 69

### Diana Fritillary
pg. 71

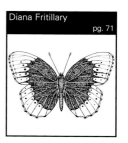

### Spicebush Swallowtail
pg. 73

### Eastern Tiger Swallowtail
pg. 75

### Palamedes Swallowtail
pg. 77

### Polydamas Swallowtail
pg. 79

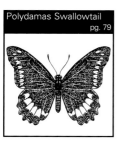

### Cassius Blue
pg. 81

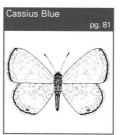

### Ceraunus Blue
pg. 83

### Eastern Tailed-Blue
pg. 85

### Spring Azure
pg. 87

### Summer Azure
pg. 89

## Silvery Blue
pg. 91

## Appalachian Azure
pg. 93

## White M Hairstreak
pg. 95

## Great Purple Hairstreak
pg. 97

## Eastern Pygmy-Blue
pg. 99

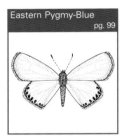

## Eastern Tailed-Blue
pg. 101

## Dusky Roadside-Skipper
pg. 103

## Brown Elfin
pg. 105

## Frosted Elfin
pg. 107

## Swarthy Skipper
pg. 109

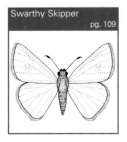

## Tawny-edged Skipper
pg. 111

## Eastern Pine Elfin
pg. 113

| Henry's Elfin pg. 115 | Pepper and Salt Skipper pg. 117 | Coral Hairstreak pg. 119 |
|---|---|---|
|  | 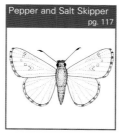 |  |

| Eufala Skipper pg. 121 | Edwards' Hairstreak pg. 123 | Hickory Hairstreak pg. 125 |
|---|---|---|
|  |  |  |

| King's Hairstreak pg. 127 | Banded Hairstreak pg. 129 | Crossline Skipper pg. 131 |
|---|---|---|
|  |  |  |

| Peck's Skipper pg. 133 | Whirlabout pg. 135 | Southern Hairstreak pg. 137 |
|---|---|---|
|  |  |  |

### Striped Hairstreak
pg. 139

### Common Roadside-Skipper
pg. 141

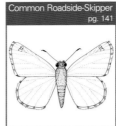

### Hayhurst's Scallopwing
pg. 143

### Salt Marsh Skipper
pg. 145

### Zabulon Skipper
pg. 147

### Northern Broken-Dash
pg. 149

### Southern Broken-Dash
pg. 151

### Little Glassywing
pg. 153

### Sachem
pg. 155

### Carolina Satyr
pg. 157

### Bell's Roadside-Skipper
pg. 159

### Carolina Roadside-Skipper
pg. 161

### Clouded Skipper
pg. 163

### Cobweb Skipper
pg. 165

### Dun Skipper
pg. 167

### Reversed Roadside-Skipper
pg. 169

### Dreamy Duskywing
pg. 171

### Twin-spot Skipper
pg. 173

### Two-spotted Skipper
pg. 175

### Lace-winged Roadside-Skipper
pg. 177

### Confused Cloudywing
pg. 179

### Southern Cloudywing
pg. 181

### Mottled Duskywing
pg. 183

### Northern Cloudywing
pg. 185

**31**

## Gemmed Satyr
pg. 187

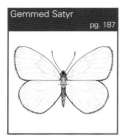

## Horace's Duskywing
pg. 189

## Wild Indigo Duskywing
pg. 191

## Dotted Skipper
pg. 193

## Hobomok Skipper
pg. 195

## Sleepy Duskywing
pg. 197

## Yehl Skipper
pg. 199

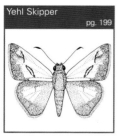

## Dusted Skipper
pg. 201

## Mitchell's Satyr
pg. 203

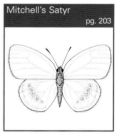

## Hoary Edge
pg. 205

## Georgia Satyr
pg. 207

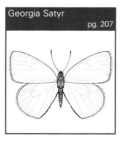

## Dorantes Skipper
pg. 209

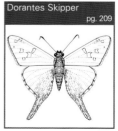

### Leonard's Skipper
pg. 211

### Ocola Skipper
pg. 213

### Little Wood Satyr
pg. 215

### Juvenal's Duskywing
pg. 217

### Zarucco Duskywing
pg. 219

### Golden-banded Skipper
pg. 221

### Long-tailed Skipper
pg. 223

### American Snout
pg. 225

### Viola's Wood Satyr
pg. 227

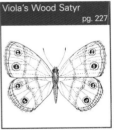

### Brazilian Skipper
pg. 229

### Appalachian Brown
pg. 231

### Silver-spotted Skipper
pg. 233

**33**

## Common Buckeye
pg. 235

## Cofaqui Giant-Skipper
pg. 237

## Northern Pearly Eye
pg. 239

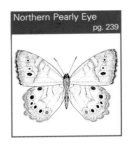

## Hackberry Butterfly
pg. 241

## Common Wood Nymph
pg. 243

## Southern Pearly Eye
pg. 245

## Tawny Emperor
pg. 247

## Yucca Giant-Skipper
pg. 249

## Creole Pearly Eye
pg. 251

## Compton Tortoiseshell
pg. 253

## Giant Swallowtail
pg. 255

## Red-banded Hairstreak
pg. 257

| Dusky Azure pg. 259 | Gray Hairstreak pg. 261 | Early Hairstreak pg. 263 |
|---|---|---|

| Hessel's Hairstreak pg. 265 | Juniper Hairstreak pg. 267 | Southern Skipperling pg. 269 |
|---|---|---|

 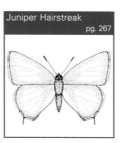 

| Little Metalmark pg. 271 | Least Skipper pg. 273 | European Skipper pg. 275 |
|---|---|---|

| Phaon Crescent pg. 277 | Fiery Skipper pg. 279 | Whirlabout pg. 281 |
|---|---|---|

American Copper
pg. 283

Harvester
pg. 285

Delaware Skipper
pg. 287

Zabulon Skipper
pg. 289

Gorgone Checkerspot
pg. 291

Sachem
pg. 293

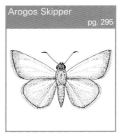

Arogos Skipper
pg. 295

Indian Skipper
pg. 297

Meske's Skipper
pg. 299

Pearl Crescent
pg. 301

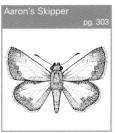

Aaron's Skipper
pg. 303

Hobomok Skipper
pg. 305

### Tawny Crescent
pg. 307

### Yehl Skipper
pg. 309

### Dion Skipper
pg. 311

### Meadow Fritillary
pg. 313

### Berry's Skipper
pg. 315

### Byssus Skipper
pg. 317

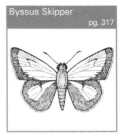

### Duke's Skipper
pg. 319

### Sleepy Orange
pg. 321

### Silvery Checkerspot
pg. 323

### Northern Crescent
pg. 325

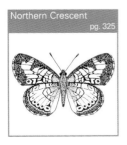

### Broad-winged Skipper
pg. 327

### Rare Skipper
pg. 329

**37**

**Palatka Skipper**
pg. 331

**Variegated Fritillary**
pg. 333

**Orange Sulphur**
pg. 335

**Painted Lady**
pg. 337

**American Painted Lady**
pg. 339

**Green Comma**
pg. 341

**Eastern Comma**
pg. 343

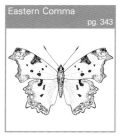

**Gray Comma**
pg. 345

**Goatweed Butterfly**
pg. 347

**Question Mark**
pg. 349

**Gulf Fritillary**
pg. 351

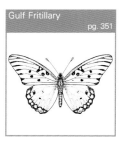

**Viceroy**
pg. 353

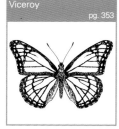

## Aphrodite Fritillary
pg. 355

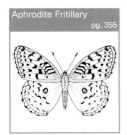

## Queen
pg. 357

## Great Spangled Fritillary
pg. 359

## Regal Fritillary
pg. 361

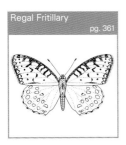

## Monarch
pg. 363

## Diana Fritillary
pg. 365

## Common/White Checkered-Skipper
pg. 367

## Tropical Checkered-Skipper
pg. 369

## Barred Sulphur
pg. 371

## West Virginia White
pg. 373

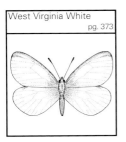

## Falcate Orangetip
pg. 375

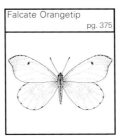

## Checkered White
pg. 377

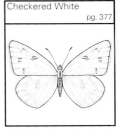

| Cabbage White | Great Southern White | White Peacock |
|---|---|---|
| pg. 379 | pg. 381 | pg. 383 |
| 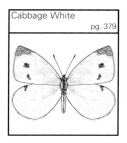 | 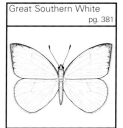 | 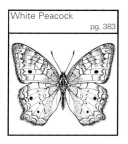 |

| Clouded Sulphur | Dainty Sulphur | Little Sulphur |
|---|---|---|
| pg. 385 | pg. 387 | pg. 389 |
|  |  |  |

| Barred Sulphur | Southern Dogface | Clouded Sulphur |
|---|---|---|
| pg. 391 | pg. 393 | pg. 395 |
|  |  |  |

| Cloudless Sulphur | Orange-barred Sulphur | Eastern Tiger Swallowtail |
|---|---|---|
| pg. 397 | pg. 399 | pg. 401 |
| 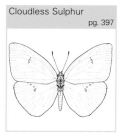 | 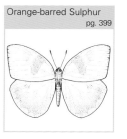 |  |

# Common Name
*Scientific Name*

color section indicators →

**Family/Subfamily:** tells which family and subfamily the butterfly belongs to (see p. 9–13 for descriptions)

**Wingspan:** gives minimum and maximum wing spans, from one forewing tip to the other

**Above:** description of upper, or dorsal, surface of wings

**Below:** description of lower, or ventral, surface of wings

**Sexes:** describes differences in appearance between male and female

**Egg:** description of eggs and where they are deposited

**Larva:** description of the butterfly's larva, or caterpillar

**Larval Host Plants:** lists plants that eggs and larva are likely to be found on

**Habitat:** describes where you're likely to find the butterflies

**Broods:** lists number of broods, or generations, hatched in a span of one year

**Abundance:** when the butterflies are flying, this tells you how often you're likely to encounter them

**Compare:** describes differences among similar-looking species

range map shows where in the Carolinas this butterfly is present

Resident: predictably present
Visitor: occasionally present
Stray: rarely present

phenogram: shows population flux throughout the year

Resident Visitor Stray

Jan. Feb. Mar. Apr. May June July Aug. Sept. Oct. Nov. Dec.

**Dorsal (above)**

**Ventral (below)**

silhouette behind Comments section shows actual average size of butterfly

illustration shows field marks and features to look for

**Comments:** The Common Sootywing indeed looks as if it fell into a pail of ashes. This solid brownish black coloring and the small white spots on the wings quickly distinguish it from other duskywings and cloudywings. The Common Sootywing is avidly drawn to available flowers. Adults have a low, erratic flight and scurry around quickly, stopping occasionally to rest on nearby vegetation or bare soil with their wings spread. The larvae construct individual shelters on the host by folding over part of a leaf with silk.

# Common Sootywing
*Pholisora catullus*

**Family/Subfamily:** Skippers (Hesperiidae)/
Spread-wing Skippers (Pyrginae)

**Wingspan:** 0.90–1.25" (2.3–3.2 cm)

**Above:** shiny dark brown to black with a variable number
of small white spots on the forewing and a few on top
of the head

**Below:** as above but paler brown

**Sexes:** similar, although female often has larger white
forewing spots

**Egg:** reddish pink, laid singly on upperside of host leaves

**Larva:** pale gray-green with a narrow dorsal stripe, pale
green lateral stripes, a black collar and black head;
body is covered with numerous tiny yellow-white dots,
each bearing a short hair

**Larval Host Plants:** Lamb's Quarters, Mexican Tea,
Spiny Amaranth

**Habitat:** open, disturbed sites including roadsides, old
fields, utility easements and fallow agricultural land

**Broods:** two or more generations

**Abundance:** occasional

**Compare:** Northern Cloudywing (pg. 185) is lighter
brown with small, glassy forewing spots, a light wing
fringe and two central dark bands across the ventral
hindwings.

Resident

| Jan. | Feb. | Mar. | Apr. | May | June | July | Aug. | Sept. | Oct. | Nov. | Dec. |

**male**

Dorsal (above)
small white spots
otherwise black

white spots on head

occasionally has a row
of small white spots
on hindwing

Ventral (below)

43

Ventral

Larva

**Comments:** This primarily Canadian butterfly is found in portions of the central Appalachians where it is considered rare and declining. It is a rare stray or possibly isolated colonist southward into the mountains of North Carolina. Superficially similar to the widespread and abundant common Checkered-Skipper, it appears noticeably darker with reduced white spotting on the wings above. Larvae pupate in late summer and overwinter in leaf shelters.

# Grizzled Skipper
*Pyrgus centaureae*

**Family/Subfamily:** Skippers (Hesperiidae)/
Spread-wing Skippers (Pyrginae)

**Wingspan:** 1.1–1.3" (2.8–3.3 cm)

**Above:** dark gray to black with scattered small, white spots and black-and-white checkered fringes; forewing has scattered white spots; central band is missing a white spot below cell-end bar

**Below:** white with irregular olive gray bands and spots

**Sexes:** similar

**Egg:** pale green, laid singly on the underside of host leaves

**Larva:** gray green with a black head

**Larval Host Plants:** Canada Cinquefoil

**Habitat:** open woodlands, barrens, forest clearings and margins, utility easements and along forest trails

**Broods:** single generation

**Abundance:** very rare stray to the Carolinas

**Compare:** Common/White Checkered-Skipper (pg. 367) is larger and has more extensive white spotting.

Stray

Jan. Feb. Mar. Apr. May June July Aug. Sept. Oct. Nov. Dec.

**Dorsal (above)**
scattered small white spots

missing white spot below cell

checkered fringes

**Ventral (below)**
irregular bands and spots

**45**

Larva

**Comments:** With a wingspan approaching two inches, the Great Purple Hairstreak significantly dwarfs most other members of its family. It typically dwells high up in the canopy close to its mistletoe host. Nonetheless, adults readily come down to feed on available flowers and can be closely observed. Easily identified by its orange abdomen, the butterfly's bold ventral coloration warns predators that it is highly distasteful. This butterfly typically rests and feeds with its wings closed, so its blue color is seen primarily in flight or when the individual is basking.

# Great Purple Hairstreak
*Atlides halesus*

**Family/Subfamily:** Gossamer Wings (Lycaenidae)/ Hairstreaks (Theclinae)

**Wingspan:** 1.0–1.7" (2.5–4.3 cm)

**Above:** male is bright metallic purple-blue with black margins and dark forewing stigma; female is dull black with metallic blue scaling limited to wing bases; two hindwing tails

**Below:** dull brownish black with metallic green and blue spots near tails and red spots at wing bases; underside of abdomen reddish orange; head and thorax have white spots

**Sexes:** dissimilar; female duller with less blue

**Egg:** green, laid singly on host leaves

**Larva:** green with numerous short hairs

**Larval Host Plants:** Mistletoe

**Habitat:** woodland edges, adjacent open areas, gardens and parks

**Broods:** multiple generations

**Abundance:** occasional

**Compare:** White M Hairstreak (pg. 95) is smaller with a narrow white M on the underside of the hindwing near the tails and prominent red spot.

Resident

Jan. Feb. Mar. Apr. May June July Aug. Sept. Oct. Nov. Dec.

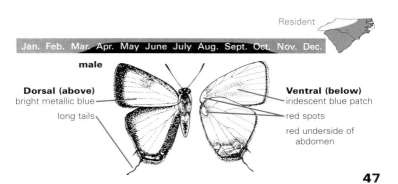

**male**

**Dorsal (above)**
bright metallic blue

long tails

**Ventral (below)**
iridescent blue patch

red spots

red underside of abdomen

**47**

Ventral    Larva

**Comments:** This distinctive dark crescent can be confused with no other butterfly. Primarily restricted to
southern and costal portions of South Carolina, it frequents the margins of watercourses and adjacent
open areas where its wetland hosts can be found.
Adults scurry low to the ground with a gliding and
somewhat erratic flight. They frequently stop to perch
on low vegetation or to nectar at available flowers. The
eastern subspecies *(P. texana seminole)*, has brighter
reddish orange basal coloration and larger cream spots
on the wings above.

# Texan Crescent
*Phyciodes texana*

**Family/Subfamily:** Brush-foots (Nymphalidae)/
True Brush-foots (Nymphalinae)

**Wingspan:** 1.3–1.8" (3.3–4.6 cm)

**Above:** black with cream white spots and reddish orange
basal markings; forewing margin is noticeably concave

**Below:** hindwing is mottled dark and light brown with a
distinct cream band through the center

**Sexes:** similar

**Egg:** pale green, laid in clusters on the underside of host
leaves

**Larva:** dark brownish black with white dashes, a cream
yellow lateral band with orange-based yellow ochre
spines, and a black head; dorsal spines black

**Larval Host Plants:** various acanthus family plants
including Carolina Wild Petunia, Branched Foldwing
and snake-herb

**Habitat:** stream and river corridors and open, disturbed
sites

**Broods:** multiple generations

**Abundance:** occasional

**Compare:** unique

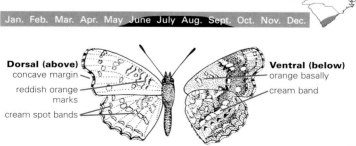

Resident

Jan. Feb. Mar. Apr. May June July Aug. Sept. Oct. Nov. Dec.

**Dorsal (above)**
concave margin
reddish orange marks
cream spot bands

**Ventral (below)**
orange basally
cream band

Ventral

Larva

**Comments:** With its outrageous color, the Baltimore is a must-see butterfly! It utilizes both primary and secondary larval hosts. Turtlehead serves as the primary host on which females lay eggs. Upon hatching, the larvae construct a communal silken web on the plant and feed gregariously inside. By the end of summer, they stop feeding, construct a new silken web near the base of the plant and overwinter. In spring they move to secondary hosts to complete development. Adults have a quick, low flight made up of rapid wing beats followed by brief periods of gliding. Both sexes readily perch on low vegetation with their wings held open

# Baltimore Checkerspot
*Euphydryas phaeton*

**Family/Subfamily:** Brush-foots (Nymphalidae)/
True Brush-foots (Nymphalinae)

**Wingspan:** 1.75–2.50" (4.4–6.6 cm)

**Above:** black with reddish orange spots along the outer
wing margins and several rows of cream white spots

**Below:** marked as above with several large reddish
orange basal spots on both wings

**Sexes:** similar

**Egg:** yellow, soon turning reddish, laid in large clusters of
several hundred on the underside of host leaves

**Larva:** tawny orange with black transverse stripe and
several rows of black, branched spines; front and rear
end segments are black

**Larval Host Plants:** Turtlehead is the primary host;
secondary hosts include a wider range of plants such
as false foxglove, Narrowleaf Plantain, Canadian
Lousewort, and Southern Arrowwood

**Habitat:** wet meadows, bogs, stream margins, marshes

**Broods:** single generation

**Abundance:** uncommon

**Compare:** unique

Resident

Jan. Feb. Mar. Apr. May June July Aug. Sept. Oct. Nov. Dec.

**Dorsal (above)**
reddish orange
marginal spots

several rows of
pale spots on
both wings

**Ventral (below)**
reddish orange
crescents

reddish orange spots

**51**

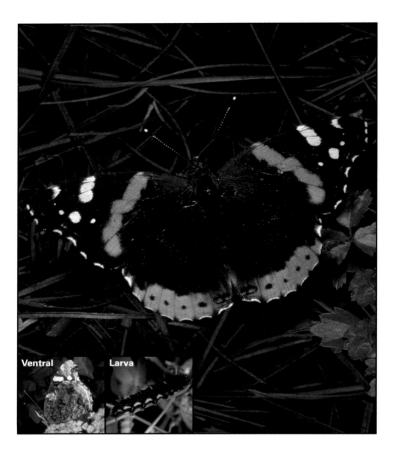

Ventral  Larva

**Comments:** The Red Admiral is a medium-sized dark butterfly that is quickly distinguished from any other species by its distinctive reddish orange forewing band. Particularly common in spring, the butterfly will readily explore suburban gardens. Adults have a quick, erratic flight and are often quite wary. Males frequently perch on low vegetation or on the ground in sunlit locations. Adults occasionally nectar at flowers, but more frequently visit sap flows, dung or fermenting fruit. The larvae construct individual shelters on the host by folding together one or more leaves with silk.

# Red Admiral
*Vanessa atalanta*

**Family/Subfamily:** Brush-foots (Nymphalidae)/
True Brush-foots (Nymphalinae)

**Wingspan:** 1.75–2.50" (4.4–6.4 cm)

**Above:** dark brownish black with reddish orange hindwing border and distinct, reddish orange median
forewing band; forewing has small white spots near
apex

**Below:** forewing as above with blue scaling and paler
markings; hindwing ornately mottled with dark brown,
blue and cream in bark-like pattern

**Sexes:** similar

**Egg:** small green eggs laid singly on host leaves

**Larva:** variable; pinkish gray to charcoal with lateral row
of cream crescent-shaped spots and numerous
branched spines

**Larval Host Plants:** False Nettle, Pellitory and nettles

**Habitat:** moist woodlands, forest edges, roadside
ditches, along canals and ponds, wetlands and gardens

**Broods:** multiple generations

**Abundance:** occasional; locally abundant

**Compare:** unique

Resident

Jan. Feb. Mar. Apr. May June July Aug. Sept. Oct. Nov. Dec.

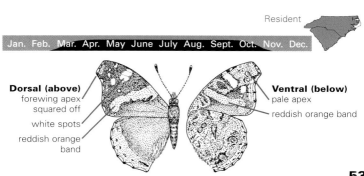

**Dorsal (above)**
forewing apex
squared off

white spots

reddish orange
band

**Ventral (below)**
pale apex

reddish orange band

**53**

Ventral

Larva

**Comments:** This colorful northern butterfly is a rare stray to the Carolinas. In years when the species has a large population outbreak, vagrant individuals may be found in the southern Appalachians.

# Milbert's Tortoiseshell
*Nymphalis milberti*

**Family/Subfamily:** Brush-foots (Nymphalidae)/ True Brush-foots (Nymphalinae)

**Wingspan:** 1.9–2.4" (4.8–6.1 cm)

**Above:** two-toned; basal portion is velvety chocolate brown with reddish orange bars in the forewing cell, outer portion has dark wing margins bordered inwardly with a wide yellow orange band; hindwing border encloses a row of small blue spots; forewing apex is extended and squared off; hindwing bears a single short, stubby tail

**Below:** strongly two-toned; basal half dark blackish brown and outer portion grayish brown with fine striations and a dark border; resembles tree bark

**Sexes:** similar

**Egg:** green, laid in clusters on host leaves

**Larva:** black with two pale lateral bands, white speckling, and several rows of branched, black spines; the ventral surface is gray green

**Larval Host Plants:** Stinging Nettle

**Habitat:** wet meadows, stream margins, pastures and other open areas near moist woodlands

**Broods:** two generations

**Abundance:** rare

**Compare:** unique

Stray

Jan. Feb. Mar. Apr. May June July Aug. Sept. Oct. Nov. Dec.

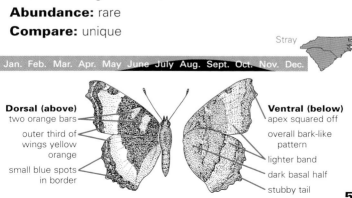

**Dorsal (above)**
- two orange bars
- outer third of wings yellow orange
- small blue spots in border

**Ventral (below)**
- apex squared off
- overall bark-like pattern
- lighter band
- dark basal half
- stubby tail

**55**

Larva

**Comments:** Impressive in size and color, the Mangrove Skipper is arguably one of the most attractive skippers in the Southeast. It is limited to coastal portions of Florida and is a very rare stray to South Carolina. Adults have a fast, powerful flight and are gone in a flash. Fortunately, they regularly stop to nectar at available flowers with their wings spread, and are easily observed. The brightly colored larvae construct individual shelters on the host by folding over sections of a leaf with silk.

# Mangrove Skipper
*Phocides pigmalion*

**Family/Subfamily:** Skippers (Hesperiidae)/ Spread-wing Skippers (Pyrginae)

**Wingspan:** 2.00–2.75" (5.1–7.0 cm)

**Above:** black with iridescent blue scaling; hindwing tapers into a small, stubby tail and has iridescent blue dashes

**Below:** black; hindwing has a row of iridescent blue dashes and basal spots

**Sexes:** similar

**Egg:** laid singly on host leaves

**Larva:** crimson with yellow bands; yellow patches on the head; larval color changes to cream-white in last instar

**Larval Host Plants:** Red Mangrove

**Habitat:** coastal mangroves and adjacent open areas

**Broods:** multiple generations

**Abundance:** extremely rare

**Compare:** unique

Stray

Jan. Feb. Mar. Apr. May June July Aug. Sept. Oct. Nov. Dec.

**male**

**Dorsal (above)**
brown-black with blue sheen

robust body

iridescent light blue spots

checkered white and black fringe

**Ventral (below)**
iridescent light blue streaks and spots

stubby tail

**57**

Ventral

Larva

**Comments:** Aptly named, the Zebra Longwing is quickly identified by the yellow-white stripes and extremely elongated wings. It is a temporary colonist of extreme southern and coastal South Carolina, and is most abundant following mild winters. It can be an abundant garden visitor. Adults have slow, graceful flight. Unlike most butterflies, the Zebra Longwing is extremely long-lived and may survive for several months. As evening approaches, they often gather on Spanish moss or overhanging branches to form a communal roost.

# Zebra Longwing
*Heliconius charitonius*

**Family/Subfamily:** Brush-foots (Nymphalidae)/ Longwing Butterflies (Heliconiinae)

**Wingspan:** 2.9–3.5" (7.4–8.9 cm)

**Above:** elongated black wings with pale yellow stripes; long, slender abdomen and antennae

**Below:** as above with small red basal spots

**Sexes:** similar

**Egg:** yellow, laid in small groups on new growth and tendrils of host; host is typically shaded or semi-shaded

**Larva:** white with small black spots and numerous long black spines

**Larval Host Plants:** Various Passion-Flowers including Maypop, Corky-Stemmed Passion-Flower, Yellow Passion-Flower and Incense Passion-Flower

**Habitat:** woodlands, forest edges and adjacent open, disturbed areas such as gardens

**Broods:** multiple generations

**Abundance:** occasional to common

**Compare:** unique

Visitor

Jan. Feb. Mar. Apr. May June July Aug. Sept. Oct. Nov. Dec.

**Dorsal (above)**
black with yellow stripes

long antennae

long abdomen

**Ventral (below)**
elongated wings

small red spots on wing base

**59**

**Ventral** **Larva**

**Comments:** The Red-Spotted Purple is one of several Carolina butterflies that mimic the toxic Pipevine Swallowtail to gain protection from predators such as birds. It is a common butterfly of immature woodlands but is rarely encountered in large numbers. Adults have a strong, gliding flight and are often quite wary. Males perch on sunlit branches along trails or forest borders and make periodic exploratory flights. Adults occasionally visit flowers but often prefer rotting fruit, dung, carrion or tree sap. Larvae eat the tip of the leaf to mid-vein and rest on the end of the vein when not actively feeding.

# Red-Spotted Purple
*Limenitis arthemis astyanax*

**Family/Subfamily:** Brush-foots (Nymphalidae)/ Admirals and Relatives (Limenitidinae)

**Wingspan:** 3.0–3.5" (7.6–8.9 cm)

**Above:** dark velvety bluish black with iridescent blue scaling on hindwing and small orange and white spots near forewing apex

**Below:** brownish black with an iridescent blue sheen and basal red-orange spots; hindwings have a row of red-orange spots toward the outer margin

**Sexes:** similar

**Egg:** gray-green, laid singly on the tips of host leaves

**Larva:** mottled green, brown and cream with two long, knobby horns on thorax; resembles a bird dropping

**Larval Host Plants:** Black Cherry, Wild Cherry and willow

**Habitat:** open woodlands, forest edges and adjacent open areas

**Broods:** multiple generations

**Abundance:** occasional

**Compare:** Pipevine Swallowtail (pg. 67) has hindwing tail and a low, rapid flight; continually flutters wings when nectaring.

Resident

Jan. Feb. Mar. Apr. May June July Aug. Sept. Oct. Nov. Dec.

**Dorsal (above)**
iridescent blue

hindwing is squared off

**Ventral (below)**
pale forewing apex

red-orange spots

Larva

**Comments:** The black and white striped Zebra Swallowtail can be confused with no other resident Eastern butterfly. Adults have a low, rapid flight and adeptly maneuver among understory vegetation. Males regularly patrol territories for females. This swallowtail has a proportionately short proboscis and is thus unable to nectar at many long, tubular flowers. It instead prefers composites and is regularly attracted to white flowers.

# Zebra Swallowtail
*Eurytides marcellus*

**Family/Subfamily:** Swallowtails (Papilionidae)/ Swallowtails (Papilioninae)

**Wingspan:** 2.5–4.0" (6.4–10.2 cm)

**Above:** white with black stripes and long, slender tails; hindwings bear a bright red patch above the eyespot; spring-forms are smaller, lighter and have shorter tails

**Below:** as above, but with a red stripe through hindwing

**Sexes:** similar

**Egg:** light green, laid singly on host leaves or budding branches

**Larva:** several color forms; may be green, green with light blue and yellow stripes or charcoal with white and yellow stripes

**Larval Host Plants:** Pawpaw

**Habitat:** rich woodlands, stream corridors, swamp margins, forest edges and old fields

**Broods:** multiple generations

**Abundance:** occasional to common

**Compare:** unique

Resident

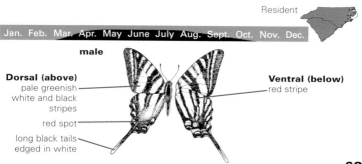

Jan. Feb. Mar. Apr. May June July Aug. Sept. Oct. Nov. Dec.

**male**

**Dorsal (above)**
pale greenish white and black stripes
red spot
long black tails edged in white

**Ventral (below)**
red stripe

**63**

Male

Female    Larva

**Comments:** The Black Swallowtail is one of our most commonly encountered garden butterflies. Its plump, green larvae, often referred to as parsley worms, feed on many cultivated herbs and may occasionally become minor nuisance pests. It is equally at home along farm roads or in rural meadows as suburban yards and urban parks. Males have a strong, rapid flight and frequently perch on low vegetation. Both sexes are exceedingly fond of flowers and readily stop to nectar. It is one of six eastern butterflies that mimic the toxic Pipevine Swallowtail to gain protection from predators.

# Black Swallowtail
*Papilio polyxenes*

**Family/Subfamily:** Swallowtails (Papilionidae)/ Swallowtails (Papilioninae)

**Wingspan:** 2.5–4.2" (6.4–10.7 cm)

**Above:** male is black with a broad, postmedian yellow band and a row of marginal yellow spots; female is mostly black with increased blue hindwing scaling and marginal yellow spots, the yellow postmedian band is reduced; both sexes have a red hindwing eyespot with a central black pupil, and a yellow-spotted abdomen

**Below:** hindwing has orange-tinted yellow spot bands

**Sexes:** dissimilar; female has reduced yellow postmedian band and increased blue hindwing scaling above

**Egg:** yellow, laid singly on host leaves

**Larva:** green with black bands and yellow-orange spots

**Larval Host Plants:** wild and cultivated members of the Carrot Family including dill, fennel and parsley

**Habitat:** old fields, roadsides, pastures, freshwater marshes, suburban gardens, agricultural land

**Broods:** multiple generations

**Abundance:** occasional to abundant

**Compare:** Spicebush Swallowtail (pg. 73) is larger, has green-blue spots and lacks black-centered hindwing eyespot. Pipevine Swallowtail (pg. 67) is mostly black and lacks yellow marking.

Resident

Jan. Feb. Mar. Apr. May June July Aug. Sept. Oct. Nov. Dec.

**male**

**Dorsal (above)**
yellow bands
blue scaling
black "pupil" in center of spot
tail

**Ventral (below)**
yellow-orange bands
faint yellow-orange cell spot

**65**

Male

Ventral

Larva

**Comments:** A relatively small member of the family, the Pipevine Swallowtail nevertheless has a strong, rapid flight. It is particularly abundant throughout the Appalachians. Adults frequently visit flowers but rarely linger at any one blossom for long. They continuously flutter their wings while feeding. The velvety black larvae sequester various toxins from their host. These chemicals render the larvae and resulting adults highly distasteful to many predators. As a result, several other butterfly species mimic the color pattern of the Pipevine Swallowtail in order to gain protection.

# Pipevine Swallowtail
*Battus philenor*

**Family/Subfamily:** Swallowtails (Papilionidae)/
Swallowtails (Papilioninae)

**Wingspan:** 2.75–4.00" (7.0–10.2 cm)

**Above:** overall black; male has iridescent greenish blue
hindwings; female is duller black with a more prominent row of white spots near the wing margins

**Below:** hindwings are iridescent blue with a row of
prominent orange spots

**Sexes:** dissimilar; female is dull black with a more prominent row of white spots

**Egg:** brownish orange, laid singly or in small clusters

**Larva:** velvety black with orange spots and numerous
long, fleshy tubercles

**Larval Host Plants:** various pipevine species, including
Dutchman's Pipe, Virginia Snakeroot and Wooly
Pipevine

**Habitat:** fields, rich woodlands, stream corridors, gardens

**Broods:** multiple generations

**Abundance:** occasional to common

**Compare:** Spicebush Swallowtail (pg. 73) is larger with
prominent crescent-shaped marginal spots. Red-
Spotted Purple (pg. 61) lacks hindwing tails. Female
Black Swallowtail (pg. 65) is larger with an orange hindwing eyespot.

Resident

Jan. Feb. Mar. Apr. May June July Aug. Sept. Oct. Nov. Dec.

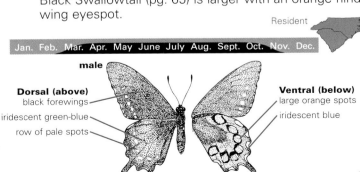

male

**Dorsal (above)**
black forewings
iridescent green-blue
row of pale spots

**Ventral (below)**
large orange spots
iridescent blue

**67**

Ventral · Larva

**Comments:** This rich beauty of a butterfly is often the first harbinger of spring. Overwintering adults occasionally become active on warm winter days and are seen flying even when snow remains on the ground. Adults emerge in early summer, aestivate until fall, then become active again to feed and build up fat reserves before seeking protected sites to hibernate— making it one of our longest-lived butterflies. It is typically encountered only in very small numbers or as solitary individuals. Adults do not visit flowers but are frequently seen at rotting fruit or sap flows.

# Mourning Cloak
*Nymphalis antiopa*

**Family/Subfamily:** Brush-foots (Nymphalidae)/
True Brush-foots (Nymphalinae)

**Wingspan:** 3.0–4.0" (7.6–10.2 cm)

**Above:** velvety black, often appearing iridescent, with
broad irregular yellow border and a row of bright purple blue spots; forewing apex is extended and squared
off; hindwing bears a single short, stubby tail

**Below:** silky black with pale wing border, heavily striated
and bark-like in appearance

**Sexes:** similar

**Egg:** light brown, laid in clusters on host leaves or twigs

**Larva:** black with a dorsal row of crimson patches, fine
white speckling and several rows of black, branched
spines

**Larval Host Plants:** birch, willow, aspen, elm and hackberry

**Habitat:** deciduous forests, clearings, riparian woodlands, woodland roads, forest edges, wetland and
watercourse margins, and adjacent open areas including suburban yards, parks and golf courses

**Broods:** single generation

**Abundance:** occasional

**Compare:** unique

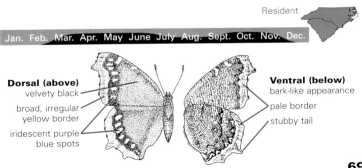

Resident

Jan. Feb. Mar. Apr. May June July Aug. Sept. Oct. Nov. Dec.

**Dorsal (above)**
velvety black

broad, irregular
yellow border

iridescent purple
blue spots

**Ventral (below)**
bark-like appearance

pale border

stubby tail

**69**

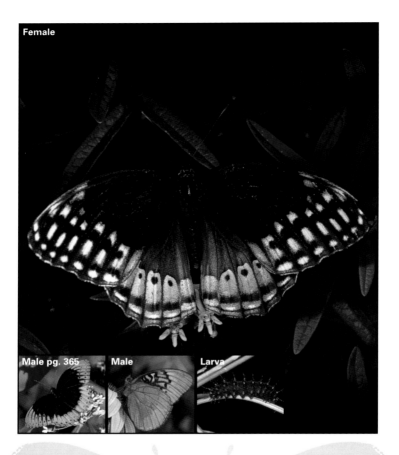

Female

Male pg. 365

Male

Larva

**Comments:** The Diana Fritillary is a showstopper! Strikingly sexually dimorphic, the blue and black female mimics the toxic Pipevine Swallowtail to presumably gain protection from predators. Although often spotty and localized, it can be fairly numerous when encountered. Adults have a strong, directed flight but readily stop to nectar at colorful wildflowers, being particularly fond of milkweeds, Joe-pye Weed and thistles. As with other fritillaries, males typically emerge several weeks before females. Young larvae overwinter and complete development the following spring.

# Diana Fritillary
*Speyeria diana*

**Family/Subfamily:** Brush-foots (Nymphalidae)/
Longwing Butterflies (Heliconiinae)

**Wingspan:** 3.5–4.4" (8.9–11.2 cm)

**Above:** male is dark unmarked blackish brown with
bright orange on the outer third; female is black basally
with white and iridescent blue spots on outer half

**Below:** forewing is orange with heavy black markings
toward base; male hindwing is brownish orange with
two rows of small narrow silver dashes; female hindwing is chocolate brown

**Sexes:** dissimilar; female is black basally with white and
iridescent blue spots on outer half; ventral hindwing is
chocolate brown

**Egg:** tiny cream eggs laid singly and somewhat haphazardly near host leaves

**Larva:** velvety black with several rows of reddish orange
based black spines

**Larval Host Plants:** various woodland violets

**Habitat:** rich, moist deciduous mountain woodlands,
stream corridors, forested roads, clearings and adjacent open areas

**Broods:** single generation

**Abundance:** occasional to locally abundant

**Compare:** unique

Resident

Jan. Feb. Mar. Apr. May June July Aug. Sept. Oct. Nov. Dec.

**male**

**Dorsal (above)**
black
orange
(female primarily black
and blue)

**Ventral (below)**
black markings toward
base

lacks prominent silvery
hindwing spots
characteristic of
other fritillaries

**71**

Male

Larva

**Comments:** The Spicebush Swallowtail is one of five
Carolina butterflies that mimic the unpalatable Pipevine
Swallowtail to gain protection from predators. Adults
are strong, agile fliers but rarely stray far from their
preferred woodland habitat and are infrequent in urban
locations. A true lover of flowers, they readily venture
out into nearby open areas in search of nectar and
continuously flutter their wings while feeding. The larvae make individual shelters by curling up both edges
of a leaf with silk. They rest motionless inside when
not actively feeding.

# Spicebush Swallowtail
*Papilio troilus*

**Family/Subfamily:** Swallowtails (Papilionidae)/ Swallowtails (Papilioninae)

**Wingspan:** 3.5–5.0" (8.9–12.7 cm)

**Above:** black with a row of large, pale greenish blue spots along the margin; hindwings have greenish blue scaling and a single, orange eyespot

**Below:** black with postmedian band of blue scaling bordered by row of yellow-orange spots on each side; abdomen black with longitudinal rows of light spots

**Sexes:** similar, female has duller hindwing scaling

**Egg:** cream, laid singly on the underside of host leaves

**Larva:** green above, reddish below with enlarged thorax, two false eyespots and several longitudinal rows of blue spots

**Larval Host Plants:** Sassafras, Red Bay and Spicebush

**Habitat:** woodlands, forest edges, wooded swamps, pastures, old fields and suburban gardens

**Broods:** multiple generations

**Abundance:** occasional

**Compare:** Pipevine Swallowtail (pg. 67), female Black Swallowtail (pg. 65) and dark-form female Eastern Tiger Swallowtail (pg. 75) all lack marginal greenish blue spots.

Resident

| Jan. | Feb. | Mar. | Apr. | May | June | July | Aug. | Sept. | Oct. | Nov. | Dec. |

**Dorsal (above)**
orange spot
large pale green spots
iridescent green-blue patch
spoon-shaped tails

**Ventral (below)**
yellow-orange spots
blue scaling

73

Dark-form female

Female

Male pg. 401

Larva

**Comments:** Easily recognized by its bold, black stripes and yellow wings, the Tiger Swallowtail is the state butterfly of South Carolina. Adults have a strong, agile flight and often soar high in the treetops. It is fond of woodlands and waterways, but is equally at home in more urban areas. A common and conspicuous garden visitor, adults favor red, pink and purple flowers. Dark-form females mimic the toxic Pipevine Swallowtail to gain protection from predators. Females exhibit numerous intermediate-colored forms.

# Eastern Tiger Swallowtail
*Papilio glaucus*

**Family/Subfamily:** Swallowtails (Papilionidae)/ Swallowtails (Papilioninae)

**Wingspan:** 3.5–5.5" (8.9–14.0 cm)

**Above:** yellow with black forewing stripes and broad black wing margins; single row of yellow spots along outer edge of each wing

**Below:** yellow with black stripes and black wing margins; hindwing margins have increased blue scaling and a single submarginal row of yellow-orange, crescent-shaped spots; abdomen yellow with black stripes

**Sexes:** dissimilar; male always yellow but females have two color forms; yellow female has increased blue scaling in black hindwing border; dark-form female is mostly black with extensive blue hindwing markings

**Egg:** green, laid singly on upper surface of host leaves

**Larva:** green; enlarged thorax and two small false eyespots

**Larval Host Plants:** Black Cherry, Wild Cherry, ash and Tulip Tree

**Habitat:** mixed forests, wooded swamps, forest edges, suburban gardens, old fields

**Broods:** multiple generations

**Abundance:** occasional to common

**Compare:** Pipevine Swallowtail (pg. 67) is much smaller. Spicebush Swallowtail (pg. 73) has greenish-blue spots on hindwing margin.

Resident

| Jan. | Feb. | Mar. | Apr. | May | June | July | Aug. | Sept. | Oct. | Nov. | Dec. |

male

**Dorsal (above)**
yellow with black stripes

wide black border

yellow spots

long tail

**Ventral (below)**
yellow-orange spots

blue scaling

Larva

**Comments:** Primarily restricted to the Atlantic Coastal Plain, the Palamedes Swallowtail is abundant in moist lowland woodlands and evergreen swamps. It may be commonly encountered in suburban gardens or along roadsides. Adults have a strong, directed flight and avidly nectar at available blooms. Males often congregate at moist ground to imbibe diluted minerals and salts. The larvae have an enlarged thorax with a prominent pair of false eyespots that resemble the head of a small lizard or snake.

# Palamedes Swallowtail
*Papilio palamedes*

**Family/Subfamily:** Swallowtails (Papilionidae)/ Swallowtails (Papilioninae)

**Wingspan:** 3.5–5.5" (8.9–14.0 cm)

**Above:** black with a broad, postmedian yellow band and a row of marginal yellow spots

**Below:** hindwing has a band of blue scaling bordered by median and submarginal rows of yellow-orange spots; narrow yellow line near wing base runs parallel to the abdomen, no other black colored swallowtail in the Carolinas has this marking

**Sexes:** similar

**Egg:** cream, laid singly on leaves; prefers new growth

**Larva:** green above, reddish below with enlarged thorax and two false eyespots

**Larval Host Plants:** Red Bay

**Habitat:** wooded swamps, hammocks, forest edges, suburban gardens, moist woodlands and evergreen swamps

**Broods:** multiple generations

**Abundance:** occasional to abundant

**Compare:** Black Swallowtail (pg. 65) is smaller and has a black-centered orange hindwing eyespot. Giant Swallowtail (pg. 255) has crossing yellow dorsal bands.

Resident

Jan. Feb. Mar. Apr. May June July Aug. Sept. Oct. Nov. Dec.

**Dorsal (above)**
wide yellow band
yellow tail edged in black

**Ventral (below)**
narrow yellow stripe
orange crescents

**77**

Larva

**Comments:** The Polydamas Swallowtail lacks the characteristic hindwing tails common to most other North American members of the family. This trait, combined with its broad, yellow wing bands, makes the species easy to identify. It is a fast and powerful flier with a preference for open areas. Primarily a tropical butterfly, it rarely strays into the Carolinas. Adults are good colonizers and readily disperse long distances in search of suitable hosts.

# Polydamas Swallowtail
*Battus polydamas*

**Family/Subfamily:** Swallowtails (Papilionidae)/ Swallowtails (Papilioninae)

**Wingspan:** 4.0–5.0" (10.2–12.7 cm)

**Above:** black with a prominent yellow band along the outer margin of wings; lacks tails

**Below:** hindwings have marginal row of narrow red spots; thorax and abdomen black with red spots

**Sexes:** similar

**Egg:** amber-brown, laid in small clusters on new growth of the host

**Larva:** robust, chocolate-brown with numerous short orange, fleshy tubercles

**Larval Host Plants:** various native and ornamental pipevine species

**Habitat:** fields, gardens, woodland edges, suburban parks, disturbed sites

**Broods:** multiple generations

**Abundance:** rare

**Compare:** Black Swallowtail (pg. 65) is smaller with two yellow spot bands above and hindwing tails.

Stray

| Jan. | Feb. | Mar. | Apr. | May | June | July | Aug. | Sept. | Oct. | Nov. | Dec. |

**Dorsal (above)**
black wings
yellow band
no tails

**Ventral (below)**
red spots
scalloped margin

**79**

Larva

**Comments:** The Cassius Blue is a rare find in the
Carolinas. While it may occasionally disperse northward from southern Florida, many of these isolated
records appear to be the result of immature stages of
the butterfly "piggybacking" on shipments of Leadwort
or Plumbago, now a popular landscape plant. I have
found small, thriving colonies at several commercial
nurseries in Georgia and South Carolina, all of which
receive regular plant shipments from the sunshine
state. It typically rests and feeds with its wings closed;
blue color is seen primarily in flight or while basking.

# Cassius Blue
*Leptotes cassius*

**Family/Subfamily:** Gossamer Wings (Lycaenidae)/
Blues (Polyommatinae)

**Wingspan:** 0.75–1.00" (1.9–2.5 cm)

**Above:** blue with thin black border and white wing
fringe; female is light blue with broad gray borders;
hindwing bears one dark marginal spot

**Below:** whitish with numerous gray bands and spots;
hindwing has two orange-rimmed black and metallic
blue eyespots

**Sexes:** dissimilar; female is brownish blue with pale
whitish blue basal scaling

**Egg:** greenish blue, laid singly on flower buds, flowers or
immature fruit of host

**Larva:** variable; green to green with pinkish red markings

**Larval Host Plants:** numerous plants including Wild
Tamarind, Leadwort, Blackbead and milk peas

**Habitat:** open, disturbed sites including roadsides,
vacant or weedy fields, coastal dunes, forest edges
and gardens

**Broods:** multiple generations

**Abundance:** rare

**Compare:** Ceraunus Blue (pg. 83) is more uniform gray
beneath with white-rimmed black spots near wing
base.

Stray

Jan. Feb. Mar. Apr. May June July Aug. Sept. Oct. Nov. Dec.

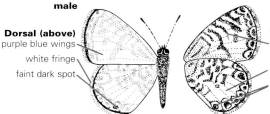

**male**

**Dorsal (above)**
purple blue wings
white fringe
faint dark spot

**Ventral (below)**
numerous dark bands
and spots
white open areas
orange-rimmed black
and metallic blue
eyespots

Male

Ventral

**Comments:** The Ceraunus Blue scurries erratically among low-growing vegetation and may be easily overlooked. Adults often rest with their wings partially open, providing a good view of their attractive dorsal coloration. Both sexes actively visit flowers. Males periodically congregate at damp ground. This butterfly typically rests and feeds with its wings closed, so its blue color is seen primarily in flight or when the butterfly is basking in the sun.

## Ceraunus Blue
*Hemiargus ceraunus*

**Family/Subfamily:** Gossamer Wings (Lycaenidae)/ Blues (Polyommatinae)

**Wingspan:** 0.75–1.00" (1.9–2.5 cm)

**Above:** male is bright lavender blue with a narrow black border and white wing fringe; single black dot along hindwing margin

**Below:** light gray with numerous dark markings and spots; hindwing has a prominent orange-rimmed black spot with metallic blue scaling

**Sexes:** dissimilar; female is brown with blue scaling at wing bases

**Egg:** greenish blue, laid singly on flower buds of host

**Larva:** highly variable; green with a red lateral stripe to highly patterned pinkish red

**Larval Host Plants:** numerous Fabaceous plants including Hairy Indigo, Creeping Indigo, Partridge Pea and milk peas

**Habitat:** open, disturbed sites including roadsides, vacant fields, utility easements and fallow agricultural land

**Broods:** multiple generations

**Abundance:** rare

**Compare:** Cassius Blue (pg. 81) is generally larger, and is chalky white beneath with numerous dark bands.

Visitor Stray

| Jan. | Feb. | Mar. | Apr. | May | June | July | Aug. | Sept. | Oct. | Nov. | Dec. |

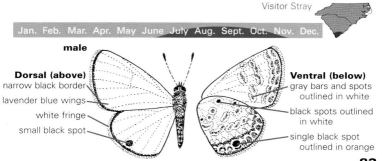

male

**Dorsal (above)**
narrow black border
lavender blue wings
white fringe
small black spot

**Ventral (below)**
gray bars and spots outlined in white
black spots outlined in white
single black spot outlined in orange

Male

Female pg. 101

Ventral

Larva

**Comments:** The Eastern Tailed-Blue is generally abundant throughout most of the eastern United States. Adults have a weak, dancing flight and readily stop to nectar. Males often gather in small clusters on damp ground. Do not rely solely on the presence of hindwing tails for identification, as they are fragile and may often be lost as part of normal wing wear. This butterfly typically rests and feeds with its wings closed, so its blue color is seen primarily in flight or when the butterfly is basking in the sun.

# Eastern Tailed-Blue
*Everes comyntas*

**Family/Subfamily:** Gossamer Wings (Lycaenidae)/
Blues (Polyommatinae)

**Wingspan:** 0.75–1.00" (1.9–2.5 cm)

**Above:** male is blue with brown border; female is
brownish gray; both sexes have one or two small
orange and black hindwing spots above single tail

**Below:** silvery-gray with numerous dark spots and
bands; hindwing has two small orange-capped black
spots above tail

**Sexes:** dissimilar; female is primarily brownish gray

**Egg:** pale green, laid singly on flowers or young leaves of
host

**Larva:** variable, typically green with dark dorsal stripe and
light lateral stripes

**Larval Host Plants:** various herbaceous Fabaceae
including clovers, bush clovers and beggarweeds

**Habitat:** open, disturbed sites including roadsides,
vacant lots and old fields

**Broods:** multiple generations

**Abundance:** occasional

**Compare:** Summer Azure (pg. 89) lacks hindwing tails.
Gray Hairstreak (pg. 261) has distinct black-and-white
(occasionally orange) ventral hindwing stripe.

Resident

Jan. Feb. Mar. Apr. May June July Aug. Sept. Oct. Nov. Dec.

male

**Dorsal (above)**
bright blue

one or two orange-
capped black spots

tail

**Ventral (below)**
black spots and bars
outlined in white

one or two orange-
capped black spots

**85**

Larva

**Comments:** Regularly encountered earlier than most other species, the Spring Azure is by far the most conspicuous early-season blue in the Carolinas, often seen even before many of the colorful spring flowering trees and shrubs are in full bloom. It may frequently wander into nearby suburban yards and gardens. Adults have a moderately slow flight and erratically scurry from ground level to canopy height moving just over the surface of the vegetation. They are extremely fond of flowers and often congregate at damp ground. It typically rests and feeds with its wings closed; blue color is seen primarily in flight or while basking.

# Spring Azure
*Celastrina ladon*

**Family/Subfamily:** Gossamer Wings (Lycaenidae)/
Blues (Polyommatinae)

**Wingspan:** 0.75–1.25" (2.0–3.2 cm)

**Above:** male is pale blue with narrow, faint dark
forewing border; lacks white scaling on wings

**Below:** somewhat variable; dusky gray with black spots
and dark scaling along hindwing margin and often a
dark patch in the center of the hindwing

**Sexes:** dissimilar; female has a broader, more extensive
dark forewing border above

**Egg:** whitish green; laid singly on flower buds of host

**Larva:** variable; green to pinkish green to whitish with
dark dorsal stripe and cream bands

**Larval Host Plants:** flowers of various trees and
shrubs including Black Cherry, blueberry, Flowering
Dogwood, viburnum and holly

**Habitat:** open, deciduous woodlands, forest edges and
trails, roadsides, brushy fields, utility easements,
wooded swamps and gardens

**Broods:** single generation

**Abundance:** occasional to common

**Compare:** All other azures are silver gray on the wings
below with reduced dark markings.

Resident

Jan. Feb. Mar. Apr. May June July Aug. Sept. Oct. Nov. Dec.

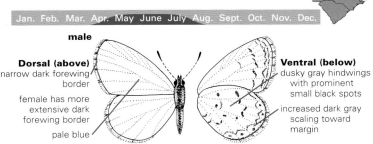

**male**

**Dorsal (above)**
narrow dark forewing
border

female has more
extensive dark
forewing border

pale blue

**Ventral (below)**
dusky gray hindwings
with prominent
small black spots

increased dark gray
scaling toward
margin

**87**

Larva

**Comments:** Now viewed as a distinct species, the Summer Azure was previously considered a lighter second generation form of the Spring Azure. Adults of this small dusty blue butterfly are found in and along woodlands but readily venture out into nearby open areas in search of nectar and frequently wander into suburban gardens. They have a moderately slow dancing flight. Unlike most other blues, they are often encountered fluttering high among the branches of trees and shrubs. Males often congregate at damp ground. It typically rests and feeds with its wings closed; blue color is seen in flight or while basking.

# Summer Azure
*Celastrina neglecta*

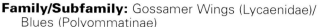

**Family/Subfamily:** Gossamer Wings (Lycaenidae)/ Blues (Polyommatinae)

**Wingspan:** 0.80–1.25" (2.0–3.2 cm)

**Above:** male is light blue with narrow, faint dark forewing border and increased white scaling on hindwing

**Below:** chalky white with small dark spots and bands

**Sexes:** dissimilar, female has increased white scaling above and wide, dark forewing borders

**Egg:** whitish green, laid singly on flower buds of host

**Larva:** variable; green to pinkish green with dark dorsal stripe and cream bands

**Larval Host Plants:** flowers of various trees and shrubs including New Jersey Tea, Wing-stem, hollies and sumac

**Habitat:** open, deciduous woodlands, forest edges, roadsides, old fields and utility easements and gardens

**Broods:** several generations

**Abundance:** occasional to common

**Compare:** Spring Azure (pg. 87) is duskier gray and more heavily marked below; lacks dorsal white scaling. Appalachian Azure (pg. 93) is slightly larger but may not reliably be separated in the field.

Resident

| Jan. | Feb. | Mar. | Apr. | May | June | July | Aug. | Sept. | Oct. | Nov. | Dec. |
|------|------|------|------|-----|------|------|------|-------|------|------|------|

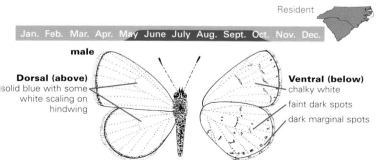

**male**

**Dorsal (above)**
solid blue with some white scaling on hindwing

**Ventral (below)**
chalky white

faint dark spots

dark marginal spots

**89**

Male

Ventral

Larva

**Comments:** The bright Silvery Blue is a beautiful spring butterfly. It is typically encountered in clearings in or near deciduous woods, often alongside several of the early-season azures. Restricted to the western Carolina mountains, the Silvery Blue typically occurs in low-density, isolated colonies, seldom far from stands of its larval hosts. Adults have a low, quick and often directed flight but frequently stop to nectar at various small flowers and are particularly fond of their host blossoms. Males often gather at mud puddles to imbibe moisture. It typically rests and feeds with its wings closed; blue color is seen in flight or while basking.

# Silvery Blue
*Glaucopsyche lygdamus*

**Family/Subfamily:** Gossamer Wings (Lycaenidae)/ Blues (Polyommatinae)

**Wingspan:** 1.00–1.25" (2.5–3.2 cm)

**Above:** male is uniform bright metallic silvery blue with narrow black wing borders and a white fringe

**Below:** light brownish gray with a prominent row of white-rimmed round black spots across the wings

**Sexes:** dissimilar; female is charcoal gray with dusky metallic blue overscaling and broad dark wing borders

**Egg:** pale blue-green, laid singly young shoots, new leaves or flower buds of host

**Larva:** gray green with a dark green dorsal stripe and white, oblique dashes; turns reddish prior to pupation

**Larval Host Plants:** Carolina Vetch, Veiny Pea and other legumes

**Habitat:** moist openings or clearings in deciduous woodlands, utility corridors, forested roads and brushy fields

**Broods:** single generation

**Abundance:** occasional

**Compare:** unique

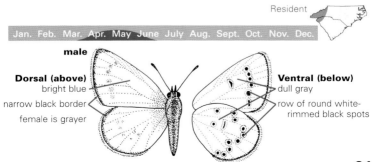

Resident

Jan. Feb. Mar. Apr. May June July Aug. Sept. Oct. Nov. Dec.

**male**

**Dorsal (above)**
bright blue

narrow black border

female is grayer

**Ventral (below)**
dull gray

row of round white-rimmed black spots

Larva

**Comments:** The Appalachian Azure is our largest resident blue. As its name suggests, it is restricted to the moist, deciduous forests of the western Carolinas. Its single spring generation is seldom on the wing for more than a few short weeks. It appears to be temporally isolated from other azures, flying after the peak of the Spring Azure and just before the emergence of the Summer Azure. It is generally rather uncommon, occurring in small, isolated and low-density colonies in close proximity to its sole larval host. Males often gather in numbers at mud puddles along stream banks or unpaved woodland roads. Blue color shows in flight.

## Appalachian Azure
*Celastrina neglectamajor*

**Family/Subfamily:** Gossamer Wings (Lycaenidae)/ Blues (Polyommatinae)

**Wingspan:** 1.1–1.4" (2.8–3.6 cm)

**Above:** male is uniform light blue with a narrow dark forewing border; lacks extensive white scaling

**Below:** chalky white with very pale, small dark spots; hindwing has faint zigzag band along margin enclosing an incomplete row of one to three prominent dark spots

**Sexes:** dissimilar; female has broad dark wing borders

**Egg:** pale green, laid singly (although often several on each plant) on flower buds of host

**Larva:** variable; yellow green to reddish brown with an often incomplete dark dorsal band and faint or absent oblique cream dashes

**Larval Host Plants:** flowers of Black Cohosh

**Habitat:** moist, cool shaded woodlands, forest trails, woodland roads and stream corridors

**Broods:** single generation

**Abundance:** occasional

**Compare:** Summer Azure (pg. 89) is smaller and has increased white scaling on dorsal hindwing, but may not reliably be separated in the field.

Resident

Jan. Feb. Mar. Apr. May June July Aug. Sept. Oct. Nov. Dec.

**male**

**Dorsal (above)**
narrow dark border

light blue

**Ventral (below)**
pale chalky white

incomplete row of
faint dark spots

Female

**Comments:** The White M Hairstreak is named for the narrow white band on the underside of the hindwing that forms the letter M. The true beauty of this species can be seen mainly during flight, when the bright iridescent blue scaling of the upper wing surfaces flashes in the sunlight. It is a comparatively large member of its family. Adults readily explore open areas adjacent to their woodland habitats for available flowers. Adults have a quick, erratic flight and can be difficult to follow. This butterfly typically rests and feeds with its wings closed.

# White M Hairstreak
*Parrhasius m-album*

**Family/Subfamily:** Gossamer Wings (Lycaenidae)/ Hairstreaks (Theclinae)

**Wingspan:** 1.0–1.5" (2.5–3.8 cm)

**Above:** male is bright iridescent blue with broad, black margins and two hindwing tails; female is dull black with blue scaling limited to wing bases

**Below:** brownish gray; hindwing has a single red eyespot above tail, white spot along leading margin, and a narrow white line forming a distinct M in middle of wing

**Sexes:** dissimilar; female duller with blue basal scaling

**Egg:** whitish, laid singly on twigs or buds of host

**Larva:** variable, dark green to mauve

**Larval Host Plants:** various oaks including Live Oak and white oak

**Habitat:** forest edges and clearings, woodland trails, oak hammocks, scrub and adjacent open areas

**Broods:** multiple generations

**Abundance:** occasional to common

**Compare:** Great Purple Hairstreak (pg. 97) is larger and lacks white M and red spot on ventral hindwing. Southern Hairstreak (pg. 137) has longer tails and an extensive orange submarginal hindwing band.

Resident

Jan. Feb. Mar. Apr. May June July Aug. Sept. Oct. Nov. Dec.

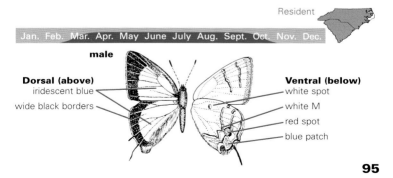

**male**

**Dorsal (above)**
iridescent blue
wide black borders

**Ventral (below)**
white spot
white M
red spot
blue patch

Larva

**Comments:** With a wingspan approaching two inches, the Great Purple Hairstreak significantly dwarfs most other members of its family. It typically dwells high up in the canopy close to its mistletoe host. Nonetheless, adults readily come down to feed on available flowers and can be closely observed. Easily identified by its orange abdomen, the butterfly's bold ventral coloration warns predators that it is highly distasteful. This butterfly typically rests and feeds with its wings closed, so its blue color is seen primarily in flight or when the individual is basking.

# Great Purple Hairstreak
*Atlides halesus*

**Family/Subfamily:** Gossamer Wings (Lycaenidae)/ Hairstreaks (Theclinae)

**Wingspan:** 1.0–1.7" (2.5–4.3 cm)

**Above:** male is bright metallic purple-blue with black margins and dark forewing stigma; female is dull black with metallic blue scaling limited to wing bases; two hindwing tails

**Below:** dull brownish black with metallic green and blue spots near tails and red spots at wing bases; underside of abdomen reddish orange; head and thorax have white spots

**Sexes:** dissimilar; female duller with less blue

**Egg:** green, laid singly on host leaves

**Larva:** green with numerous short hairs

**Larval Host Plants:** Mistletoe

**Habitat:** woodland edges, adjacent open areas, gardens and parks

**Broods:** multiple generations

**Abundance:** occasional

**Compare:** White M Hairstreak (pg. 95) is smaller with a narrow white M on the underside of the hindwing near the tails and prominent red spot.

Resident

Jan. Feb. Mar. Apr. May June July Aug. Sept. Oct. Nov. Dec.

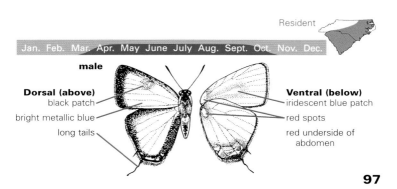

**male**

**Dorsal (above)**
black patch
bright metallic blue
long tails

**Ventral (below)**
iridescent blue patch
red spots
red underside of abdomen

Larva

**Comments:** The dainty Eastern Pygmy-Blue is our smallest butterfly. It is found only along coastal portions of southern South Carolina. Adults have a low, weak flight and scuttle just above the surface of their fleshy host plants. Nonetheless, their diminutive size often makes individuals a challenge to follow when on the wing. Easily identified, it is the only blue in the state with a row of four silvery-black spots on the hindwings below. Although occasionally abundant in certain locations, individual colonies are often sporadic and highly localized.

# Eastern Pygmy-Blue
*Brephidium isophthalma*

**Family/Subfamily:** Gossamer Wings (Lycaenidae)/ Blues (Polyommatinae)

**Wingspan:** 0.50–0.75" (1.3–1.9 cm)

**Above:** brown with some blue basal scaling and a row of dark spots along the outer margin of the hindwing

**Below:** brown with numerous white streaks and bands; hindwing margin has row of black spots with silver highlights

**Sexes:** similar

**Egg:** greenish blue, laid singly on all parts of host

**Larva:** green with dark head

**Larval Host Plants:** Saltwort and glassworts

**Habitat:** salt marshes and adjacent coastal areas

**Broods:** multiple generations

**Abundance:** occasional; locally abundant

**Compare:** Ceraunus Blue (pg. 83) is larger, gray beneath with prominent white-rimmed black spots on base of hindwing. Cassius Blue (pg. 81) is larger, strongly banded beneath with two black hindwing eyespots.

Resident

| Jan. | Feb. | Mar. | Apr. | May | June | July | Aug. | Sept. | Oct. | Nov. | Dec. |

**Dorsal (above)**
dull brown wings
brown fringe
faint blue basal scaling
row of small black spots

**Ventral (below)**
brown fringe
four prominent black spots

**99**

Female

Male pg. 85 | Ventral | Larva

**Comments:** The Eastern Tailed-Blue is generally abundant throughout most of the eastern United States. Adults have a weak, dancing flight and readily stop to nectar. Males often gather in small clusters on damp ground. Do not rely solely on the presence of hindwing tails for identification, as they are fragile and may often be lost as part of normal wing wear. This butterfly typically rests and feeds with its wings closed.

# Eastern Tailed-Blue
*Everes comyntas*

**Family/Subfamily:** Gossamer Wings (Lycaenidae)/ Blues (Polyommatinae)

**Wingspan:** 0.75–1.00" (1.9–2.5 cm)

**Above:** male is blue with brown border; female is brownish gray; both sexes have one or two small orange and black hindwing spots above single tail

**Below:** silvery-gray with numerous dark spots and bands; hindwing has two small orange-capped black spots above tail

**Sexes:** dissimilar; female is primarily brownish gray

**Egg:** pale green, laid singly on flowers or young leaves of host

**Larva:** variable, typically green with dark dorsal stripe and light lateral stripes

**Larval Host Plants:** various herbaceous Fabaceae including clovers, bush clovers and beggarweeds

**Habitat:** open, disturbed sites including roadsides, vacant lots and old fields

**Broods:** multiple generations

**Abundance:** occasional

**Compare:** Summer Azure (pg. 89) lacks hindwing tails. Gray Hairstreak (pg. 261) has distinct black-and-white (occasionally orange) ventral hindwing stripe.

Resident

Jan. Feb. Mar. Apr. May June July Aug. Sept. Oct. Nov. Dec.

male

**Dorsal (above)**
bright blue

one or two orange-capped black spots

tail

**Ventral (below)**
black spots and bars outlined in white

one or two orange-capped black spots

**101**

Larva

**Comments:** This small drab skipper is an inhabitant of open pine woodlands. Superficially similar to the more widespread and abundant Common Roadside-Skipper, it lacks the more extensive ventral gray overscaling and heavily checkered wing fringes. Rare throughout its limited range, much is yet to be learned about this reclusive, poorly understood butterfly.

# Dusky Roadside-Skipper
*Amblyscirtes alternata*

**Family/Subfamily:** Skippers (Hesperiidae)/
Banded Skippers (Hesperiinae)

**Wingspan:** 0.85–1.00" (2.2–2.5 cm)

**Above:** primarily dark blackish brown, often with a few
small white spots near the forewing apex

**Below:** dark grayish black with faint gray dusting and
checkered fringes

**Sexes:** similar

**Egg:** currently undocumented

**Larva:** blue green with faint dark dorsal stripe

**Larval Host Plants:** various grasses

**Habitat:** open pine woodlands

**Broods:** two generations

**Abundance:** occasional

**Compare:** Common Roadside-Skipper (pg. 141) has
increased gray frosting below and more prominently
checkered wing fringes.

Resident

Jan. Feb. Mar. Apr. May June July Aug. Sept. Oct. Nov. Dec.

**Dorsal (above)**
pointed forewing
small white spots
lightly checkered
fringes
dull brown

**Ventral (below)**
pale spots
dull brown with light
gray dusting
lightly checkered
fringes

**103**

Larva

**Comments:** This small but richly colored, tailless butterfly has only a single brief spring flight. Typically spotty
and highly localized in occurrence, it seldom wanders
far from stands of its larval hosts. Nonetheless, it
remains one of our most commonly encountered
elfins. The adults generally remain close to the ground,
often perching on low vegetation or resting on the
ground itself. If disturbed, they rapidly dart off but
travel only a short distance before alighting once more.
They nectar from a variety of early season flowers
including host blossoms.

# Brown Elfin
*Callophrys augustinus*

**Family/Subfamily:** Gossamer Wings (Lycaenidae)/ Hairstreaks (Theclinae)

**Wingspan:** 0.8–1.1" (2.0–2.8 cm)

**Above:** dark brown; male has dark forewing stigma

**Below:** forewing brown; hindwing dark brown at base with outer portion lighter reddish brown to mahogany

**Sexes:** similar

**Egg:** whitish, laid singly host flower buds

**Larva:** yellow green with pale yellow oblique dorsal dashes and a yellow lateral stripe

**Larval Host Plants:** primarily plants in the heath family including Black Huckleberry, Leatherleaf, Highbush Blueberry, and Blue Ridge Blueberry

**Habitat:** open woodlands, forest margins and clearings, scrub and bogs

**Broods:** single generation

**Abundance:** occasional to common

**Compare:** Eastern Pine Elfin (pg. 113) has strongly patterned hindwings with numerous dark bands outlined in white. Frosted Elfin's (pg. 107) hindwing has extensive frosting and a short, stubby tail. Henry's Elfin (pg. 115) has some white on outer portion of dark basal hindwing patch, frosting along margin of hindwing and a short, stubby tail.

Resident

| Jan. | Feb. | Mar. | Apr. | May | June | July | Aug. | Sept. | Oct. | Nov. | Dec. |
|------|------|------|------|-----|------|------|------|-------|------|------|------|

**Dorsal (above)**
brown wings
lobed anal angle of hindwing

**Ventral (below)**
hindwing much darker at base
reddish brown toward outer margin

Larva

**Comments:** Living up to its name, the Frosted Elfin
looks as if it were dusted lightly with powdered sugar
on it's ventral hindwings. Like most elfins, it tends to
be rather uncommon and highly localized, most often
encountered close to its leguminous larval hosts.
Despite its small size, generally slow, low flight and
close host association, the butterfly is considered an
effective colonizer and able to disperse some distance
to establish new populations in suitable habitat areas.
Larvae feed on host flowers and developing seed
pods.

# Frosted Elfin
*Callophrys irus*

**Family/Subfamily:** Gossamer Wings (Lycaenidae)/ Hairstreaks (Theclinae)

**Wingspan:** 0.8–1.1" (2.0–2.8 cm)

**Above:** unmarked dark brown; male has dark forewing stigma

**Below:** forewing brown; hindwing dark brown at wing base, outer portion somewhat lighter with extensive gray frosting along outer margin; frosted area contains a distinct single black spot near the short, stubby tail

**Sexes:** similar, although female lacks forewing stigma

**Egg:** whitish, laid singly host flower buds

**Larva:** blue-green with pale white oblique dorsal dashes and a pale white lateral stripe

**Larval Host Plants:** Wild Lupine and Wild Indigo

**Habitat:** open woodlands, forest margins and clearings, open, brushy fields, roadsides and scrub

**Broods:** single generation

**Abundance:** rare to occasional

**Compare:** Eastern Pine Elfin (pg. 113) has strongly patterned hindwings with numerous dark bands outlined in white. Henry's Elfin (pg. 115) lacks black spot near short, stubby tail. Brown Elfin (pg. 105) lacks hindwing frosting and short, stubby tail.

Resident

Jan. Feb. Mar. Apr. May June July Aug. Sept. Oct. Nov. Dec.

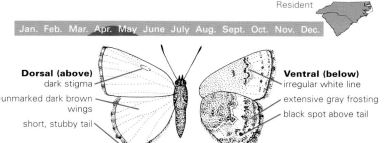

**Dorsal (above)**
dark stigma
unmarked dark brown wings
short, stubby tail

**Ventral (below)**
irregular white line
extensive gray frosting
black spot above tail

Ventral  Larva

**Comments:** This drab brown skipper is distinctly plain.
The Swarthy Skipper is reliably identified in the field by
its unmarked ventral wings with light colored veins.
However, its small size and uneventful appearance
make it an easy butterfly to quickly overlook.
Frequently encountered in open grassy sites, colonies
may be somewhat small and localized near patches of
its larval host. Like most small skippers, it has a very
quick, erratic flight but may be observed closely when
nectaring at a variety of low-growing flowers. Larvae
construct individual rolled leaf shelters on the host.

# Swarthy Skipper
*Nastra lherminier*

**Family/Subfamily:** Skippers (Hesperiidae)/
Banded Skippers (Hesperiinae)

**Wingspan:** 0.9–1.1" (2.3–2.8 cm)

**Above:** dull dark brown, occasionally with small faint forewing spots

**Below:** olive brown to yellow brown with light veins

**Sexes:** similar

**Egg:** white, laid singly on host leaves

**Larva:** elongate; pale green with a darker green dorsal stripe, a pale lateral stripe and a reddish brown head marked with vertical cream bands

**Larval Host Plants:** Little Bluestem and Kentucky Bluegrass

**Habitat:** old fields, roadsides, meadows, open, grassy areas, abandoned lots and gardens

**Broods:** two generations

**Abundance:** occasional to common

**Compare:** unique

Resident

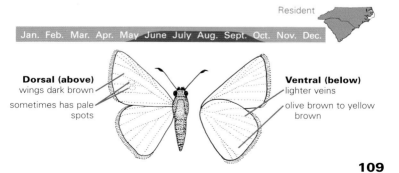

Jan. Feb. Mar. Apr. May June July Aug. Sept. Oct. Nov. Dec.

**Dorsal (above)**
wings dark brown

sometimes has pale spots

**Ventral (below)**
lighter veins

olive brown to yellow brown

**109**

Male

Male

**Comments:** The Tawny-edged Skipper's distinct orange band along the leading margin of the forewing is visible from both the dorsal and ventral surfaces. It is superficially quite similar to the Crossline Skipper. A common species, it may be frequently encountered in suburban locations including home gardens. Adults have a low, rapid flight and often alight on bare soil or low vegetation.

# Tawny-edged Skipper
*Polites themistocles*

**Family/Subfamily:** Skippers (Hesperiidae)/
Banded Skippers (Hesperiinae)

**Wingspan:** 0.8–1.2" (2.0–3.0 cm)

**Above:** male is dark brown with prominent black stigma
and tawny orange scaling along forewing costa;
female is dark brown with small yellow spots across
forewing and reduced orange along costal margin

**Below:** light brown to olive brown with distinct contrasting orange scaling along costal margin of forewing

**Sexes:** dissimilar; female darker with reduced orange coloration

**Egg:** greenish white, laid singly on host leaves

**Larva:** reddish brown with dark dorsal stripe and black
head

**Larval Host Plants:** various grasses including panic
grass, Slender Crabgrass, mannagrass and bluegrass

**Habitat:** stream corridors, wet meadows, fields, pastures and suburban yards

**Broods:** two generations

**Abundance:** occasional to common

**Compare:** Crossline Skipper (pg. 131) is larger, and usually has a faint band of small, pale spots through the
center of the yellow-brown ventral hindwing. Tolerates
drier habitats.

Resident

Jan. Feb. Mar. Apr. May June July Aug. Sept. Oct. Nov. Dec.

**male**

**Dorsal (above)**
orange along costal
margin

sinuous black stigma

**Ventral (below)**
orange scaling along
costal margin

unmarked olive to
brassy brown
hindwing

111

Larva

**Comments:** With its boldly patterned hindwings, this beautiful little spring species is our most distinctive elfin. Inhabiting a variety of open and semi-open landscapes that support stands of hard pines, the butterfly is commonly found in close association with younger trees. Adults spend much of their time perched on host branches, often high above the ground, but frequently venture down to nectar at nearby blossoms or sip moisture at damp soil.

# Eastern Pine Elfin
*Callophrys irus*

**Family/Subfamily:** Gossamer Wings (Lycaenidae)/ Hairstreaks (Theclinae)

**Wingspan:** 0.80–1.25" (2.0–3.2 cm)

**Above:** unmarked dark brown; male has a pale gray forewing stigma

**Below:** brown, strongly banded with black, reddish brown and gray; hindwing has gray marginal band

**Sexes:** similar, although female is tawnier above and lacks pale forewing stigma

**Egg:** pale green, laid singly at the base of host needles

**Larva:** bright green with cream longitudinal stripes

**Larval Host Plants:** various hard pines including Virginia Pine, Loblolly Pine, Pitch Pine and Shortleaf Pine; also Eastern White Pine

**Habitat:** open woodlands, forest margins and clearings, open, brushy fields, roadsides, scrub and utility easements

**Broods:** single generation

**Abundance:** occasional to common

**Compare:** All other elfins lack the strongly patterned ventral hindwing.

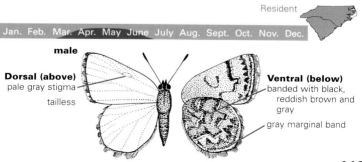

Resident

Jan. Feb. Mar. Apr. May June July Aug. Sept. Oct. Nov. Dec.

male

**Dorsal (above)**
pale gray stigma

tailless

**Ventral (below)**
banded with black, reddish brown and gray

gray marginal band

**113**

Larva (green)  Larva (red)

**Comments:** Henry's Elfin is a small, primarily brown butterfly with delicate markings and a stubby tail. Although it is widespread in the Southeast, it is generally rare and encountered only in small, localized colonies. Found in variety of semi-open areas, this spring species rarely strays far from stands of its larval host. Adults have a fast, erratic flight but regularly perch on the tips of tree branches or on shrubby vegetation. Males may be encountered sipping moisture at damp earth.

# Henry's Elfin
*Callophrys henrici*

**Family/Subfamily:** Gossamer Wings (Lycaenidae)/ Hairstreaks (Theclinae)

**Wingspan:** 0.9–1.2" (2.3–3.0 cm)

**Above:** brown with amber scaling along hindwing margin; hindwing has short, stubby tail

**Below:** brown; hindwing distinctly two-toned with dark brown basal half and light brown outer half; gray frosting along outer margin

**Sexes:** similar

**Egg:** whitish, laid singly on host twigs or flower buds

**Larva:** variable; green to reddish with oblique white dorsal markings

**Larval Host Plants:** Redbud, Dahoon Holly, American Holly and Yaupon Holly

**Habitat:** deciduous woodlands, forest edges and clearings, shrubby areas, old fields and roadsides

**Broods:** single spring generation

**Abundance:** occasional; localized

**Compare:** Eastern Pine Elfin (pg. 113) has strongly patterned ventral hindwing with numerous dark bands outlined in white. Frosted Elfin (pg. 107) has more frosting on hindwing and small dark spot near tail. Brown Elfin (pg. 105) lacks tail and gray frosting.

Resident

Jan. Feb. Mar. Apr. May June July Aug. Sept. Oct. Nov. Dec.

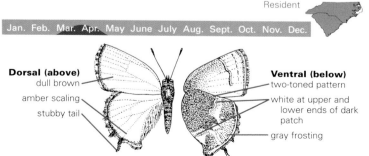

**Dorsal (above)**
dull brown
amber scaling
stubby tail

**Ventral (below)**
two-toned pattern
white at upper and lower ends of dark patch
gray frosting

**115**

**Dorsal**

**Larva**

**Comments:** Well named, this distinctive skipper does indeed look as if it was sprinkled with ample amounts of salt and pepper. Widespread throughout the east, it is a butterfly of wet habitats preferring moist meadows, forest glades, and stream corridors. Seldom common, it is most often encountered in localized colonies. Males may frequently be seen puddling at moist ground. The adults are not strongly restricted to wet sites and often wander extensively into nearby habitats in search of nectar. Larvae overwinter.

# Pepper and Salt Skipper
*Amblyscirtes hegon*

**Family/Subfamily:** Skippers (Hesperiidae)/
Banded Skippers (Hesperiinae)

**Wingspan:** 0.9–1.2" (2.3–3.0 cm)

**Above:** dark brown with checkered fringes and a band of
small white spots on the forewing

**Below:** variable; hindwing is greenish gray with cream
spot band; wing fringes are strongly checkered

**Sexes:** similar

**Egg:** light green, laid singly on host leaves

**Larva:** whitish green with a dark green dorsal line, a
paler green subdorsal line and a reddish brown head
with a pale brown crescent on each side

**Larval Host Plants:** various grasses including Fowl
Mannagrass, Indian Grass, Kentucky Bluegrass and
Indian Woodoats

**Habitat:** woodland margins, wet meadows, stream margins

**Broods:** single generation

**Abundance:** occasional

**Compare:** Bell's Roadside-Skipper (pg. 159) is generally
darker below with reduced white overscaling.

Resident

Jan. Feb. Mar. Apr. May June July Aug. Sept. Oct. Nov. Dec.

**Dorsal (above)**
white spots
brown

**Ventral (below)**
cream spot band
hindwing frosted with
light greenish gray
strongly checkered
fringes

**117**

**Comments:** This distinctive tailless hairstreak can be confused with no other small Carolina butterfly. It frequents a variety of semi-open, brushy habitats in close association with its somewhat weedy, thicket-forming hosts. Although widespread, the Coral Hairstreak is often quite localized in occurrence and is seldom encountered in large numbers. Adults have a quick, erratic flight and readily perch on the top of small trees or shrubs. Both sexes frequently visit flowers and are exceedingly fond of milkweed blossoms.

## Coral Hairstreak
*Satyrium titus*

**Family/Subfamily:** Gossamer Wings (Lycaenidae)/
Hairstreaks (Theclinae)

**Wingspan:** 0.90–1.25" (2.3–3.2 cm)

**Above:** unmarked brown; male has a small gray
forewing stigma and somewhat triangular wings

**Below:** light gray brown with a row of small, white-
rimmed black spots across both wings and a second
row of larger bright coral spots along the hindwing
margin; tailless

**Sexes:** similar, although female has more rounded wings
and lacks forewing stigma

**Egg:** cream, laid singly on host twigs, low on the trunks
of small host trees or occasionally on leaf litter below
the host

**Larva:** yellow green with pinkish red patches on each end

**Larval Host Plants:** Black Cherry, Wild Cherry,
American Plum and Chickasaw Plum

**Habitat:** overgrown fields near forest margins, brushy
woodland clearings, shrubby roadsides and trails, and
unmanaged pastures or fencerows

**Broods:** single generation

**Abundance:** uncommon to common

**Compare:** unique

Resident

| Jan. | Feb. | Mar. | Apr. | May | June | July | Aug. | Sept. | Oct. | Nov. | Dec. |

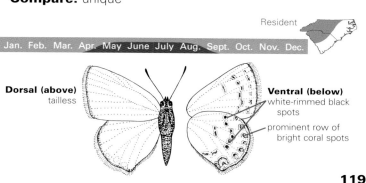

**Dorsal (above)**
tailless

**Ventral (below)**
white-rimmed black
spots

prominent row of
bright coral spots

Larva

**Comments:** The underside of the Eufala Skipper's wings appear distinctly washed-out. A year-round resident of the warm Gulf Coast and Florida, it regularly establishes temporary colonies in the Carolinas, but is unable to survive prolonged exposure to freezing temperatures. Adults avidly visit available flowers and frequent home gardens. They have a rapid, erratic flight but pause regularly to perch on low vegetation where they are easy to observe.

# Eufala Skipper
*Lerodea eufala*

**Family/Subfamily:** Skippers (Hesperiidae)/
Banded Skippers (Hesperiinae)

**Wingspan:** 0.90–1.25" (2.3–3.2 cm)

**Above:** grayish brown with several tiny semitransparent
spots on forewing; subapical spots form small, straight
band

**Below:** light brown to tan, often with faint, light hindwing
spots

**Sexes:** similar

**Egg:** cream, laid singly on or near host leaves

**Larva:** green with green and yellow stripes; head is
brownish orange below and cream above

**Larval Host Plants:** various grasses, including
Bermuda Grass, Barnyard Grass and Hooked
Bristlegrass

**Habitat:** open, grassy areas including old fields, roadsides, vacant lots, pineland edges and clearings, and
utility easements

**Broods:** two generations

**Abundance:** occasional

**Compare:** Swarthy Skipper (pg. 109) has a yellow brown
ventral hindwing with light veins.

Visitor

Jan. Feb. Mar. Apr. May June July Aug. Sept. Oct. Nov. Dec.

**male**

**Dorsal (above)**
small
semitransparent
spots

**Ventral (below)**
washed-out gray
brown, occasionally
faintly spotted

Larva

**Comments:** Edwards' Hairstreak tends to occur in highly localized colonies. Its relatively short flight period and close resemblance to other, more abundant hairstreaks makes it an easy butterfly to overlook. The developing slug-like larvae are regularly tended (and thus guarded from many predators) by ants who in turn use the larva's sugary honeydew secretions for food. Depending on the degree of this association, the butterfly may rely heavily on the ant species for survival, which may contribute to its spotty, localized distribution.

# Edwards' Hairstreak
*Satyrium edwardsii*

**Family/Subfamily:** Gossamer Wings (Lycaenidae)/ Hairstreaks (Theclinae)

**Wingspan:** 1.00–1.25" (2.5–3.2 cm)

**Above:** unmarked brown with a small orange spot near short tail; male has small dark forewing stigma

**Below:** light gray brown with a row of small, white-rimmed black spots; hindwing has a large blue patch and a series of orange spots near tails

**Sexes:** similar, although female has slightly more rounded wings and lacks dark forewing stigma above

**Egg:** cream pink, laid singly on host twigs near buds; eggs overwinter

**Larva:** dark brown with dark dorsal band and a series of pale oblique dashes along the sides

**Larval Host Plants:** various oaks including Bear Oak, Black Oak, Blackjack Oak and White Oak

**Habitat:** oak woodlands, thickets and scrub, forest margins, roadsides, utility easements and trail edges

**Broods:** single generation

**Abundance:** uncommon to common

**Compare:** Striped Hairstreak (pg. 139), Kings's Hairstreak (pg. 127), and Banded Hairstreak (pg. 129) all have bands of white-edged dashes, not spots on ventral wing surface.

Resident

Jan. Feb. Mar. Apr. May June July Aug. Sept. Oct. Nov. Dec.

**Dorsal (above)**
unmarked brown

**Ventral (below)**
white-rimmed black spots

orange spots

blue patch; often has narrow orange cap

**123**

**Comments:** The range of this rare and reclusive hairstreak just barely crosses over into the western edge of the Carolinas. A butterfly of mixed deciduous forests and second growth woodlands, it typically exists in highly localized and spotty colonies. It may further be overlooked because of its close resemblance the more widespread and common Banded Hairstreak. The adults often perch high on the branches of their hosts and occasionally dart out with a fast, erratic flight before alighting again.

# Hickory Hairstreak
*Satyrium caryaevorum*

**Family/Subfamily:** Gossamer Wings (Lycaenidae)/ Hairstreaks (Theclinae)

**Wingspan:** 1.00–1.25" (2.5–3.2 cm)

**Above:** uniform dark brown with a short hindwing tail; male has a gray forewing stigma

**Below:** brown with a row of fairly wide and somewhat offset dark dashes edged in white across both wings, a large blue hindwing patch and orange-capped black spot near the tail

**Sexes:** similar, although female lacks forewing stigma

**Egg:** pinkish brown, laid singly on twigs; eggs overwinter

**Larva:** yellow green, often with dark green dorsal stripe, yellow lateral stripe and dark dashes edged in white

**Larval Host Plants:** various hickories, walnuts and oaks including Pignut Hickory, Shagbark Hickory, Bitternut Hickory, Northern Red Oak and Bitternut

**Habitat:** mixed deciduous forests, clearings, roads and margins, and adjacent open, often brushy, areas

**Broods:** single generation

**Abundance:** rare to occasional

**Compare:** Banded Hairstreak (pg. 129) lacks white edging on inner side of dark bands. Difficult to reliably separate in the field.

Resident

| Jan. | Feb. | Mar. | Apr. | May | June | July | Aug. | Sept. | Oct. | Nov. | Dec. |

**Dorsal (above)**
dark brown

**Ventral (below)**
dark band edged in white on both sides

blue patch extends far inward

Larva

**Comments:** This lovely hairstreak is quite rare throughout its range and should be considered a "good find"
when encountered. Restricted to the shady confines
of hardwood forests and wooded swamps, it occurs in
highly localized populations in close association with
its sole larval host. Due to its reclusive nature, the
King's Hairstreak remains one of our more poorly
known and studied butterflies. Adults frequently perch
on broad sunlit leaves but seldom venture into nearby
open landscapes. Young larvae bore into host buds and
feed on developing leaves.

# King's Hairstreak
*Satyrium kingi*

**Family/Subfamily:** Gossamer Wings (Lycaenidae)/ Hairstreaks (Theclinae)

**Wingspan:** 1.00–1.25" (2.5–3.2 cm)

**Above:** uniform dark brown with two short hindwing tails

**Below:** gray brown with a row of prominent dark dashes edged in white across both wings, a large blue hindwing patch capped in orange, an orange-capped black spot near the tail, and iridescent wing fringes; freshly emerged individuals have a subtle violet blue cast

**Sexes:** similar

**Egg:** laid singly on host twigs; eggs overwinter

**Larva:** green with darker green oblique lines

**Larval Host Plants:** Common Sweetleaf

**Habitat:** hardwood forests, stream margins and wooded swamps

**Broods:** single generation

**Abundance:** rare

**Compare:** Striped Hairstreak (pg. 139) has more extensive, wider ventral bands. Banded Hairstreak (pg. 129) is more abundant, and lacks an orange cap on the ventral blue hindwing patch. Hickory Hairstreak (pg. 125) lacks an orange cap on the ventral blue hindwing patch.

Resident

Jan. Feb. Mar. Apr. May June July Aug. Sept. Oct. Nov. Dec.

**Dorsal (above)**
indented margin above tail

**Ventral (below)**
central spot offset from surrounding band

orange-capped blue patch

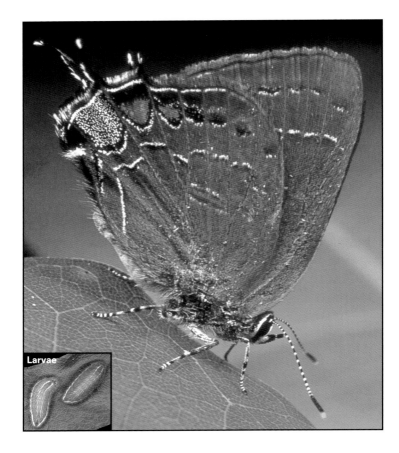

Larvae

**Comments:** A small, inconspicuous butterfly, the Banded Hairstreak is closely associated with mixed hardwood forests and can frequently be found along woodland edges or in sunlit clearings. Adults frequently visit flowers and are particularly fond of white sweet clover. Males perch on shrubs or low, overhanging limbs and aggressively defend established territories by engaging passing intruders.

# Banded Hairstreak
*Satyrium calanus*

**Family/Subfamily:** Gossamer Wings (Lycaenidae)/ Hairstreaks (Theclinae)

**Wingspan:** 1.00–1.25" (2.5–3.2 cm)

**Above:** unmarked dark brown with two hindwing tails

**Below:** variable; brown to slate gray; hindwing has dark postmedian band outlined on outer side with white and a red-capped black spot and blue patch near tails

**Sexes:** similar

**Egg:** pinkish brown, laid singly on twigs of host

**Larva:** variable, green to grayish brown with a light lateral stripe

**Larval Host Plants:** various oaks, hickories and walnuts including White Oak, Northern Red Oak, Turkey Oak, Southern Red Oak, Pignut Hickory, Shagbark Hickory, Black Walnut and Bitternut

**Habitat:** mixed deciduous forests, oak woodlands, forest clearings, roads and adjacent open areas

**Broods:** single generation

**Abundance:** occasional to common

**Compare:** Striped Hairstreak (pg. 139) has wider, more extensive bands beneath and an orange-capped blue patch near tail. Hickory Hairstreak (pg. 125) has wider, more offset dark ventral bands edged in white on both sides.

Resident

Jan. Feb. Mar. Apr. May June July Aug. Sept. Oct. Nov. Dec.

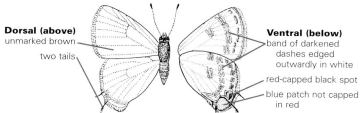

**Dorsal (above)**
unmarked brown

two tails

**Ventral (below)**
band of darkened dashes edged outwardly in white

red-capped black spot

blue patch not capped in red

**129**

**Comments:** Although frequently confused with the similar Tawny-edged Skipper, the Crossline Skipper can often be readily distinguished from its relative on the basis of habitat, preferring drier, open areas. The species may be frequently encountered in more suburban locations including home gardens and lawns where the adults can be observed nectaring at a variety of flowers. Adults have a low, rapid flight and often alight on bare soil or low vegetation.

# Crossline Skipper
*Polites origenes*

**Family/Subfamily:** Skippers (Hesperiidae)/
Banded Skippers (Hesperiinae)

**Wingspan:** 1.00–1.25" (2.5–3.2 cm)

**Above:** male is dark olive brown; forewing has tawny
orange scaling along costal margin and yellow orange
spots along the outer edge of black stigma

**Below:** hindwing is yellow brown with a faint, straight
band of small pale spots; forewing has dull tawny
orange scaling along costal margin, not strongly contrasting with color of hindwing

**Sexes:** dissimilar; female darker above with cream spots
on forewing and reduced orange along costal margin

**Egg:** greenish, laid singly on host leaves

**Larva:** dark brown with faint white mottling and a round
black head

**Larval Host Plants:** various grasses including Little
Bluestem, Purpletop Grass, mannagrass and bluegrass

**Habitat:** dry, grassy areas including old fields, woodland
meadows, pastures, forest clearings and easements

**Broods:** two generations

**Abundance:** occasional to common

**Compare:** Tawny-edged Skipper (pg. 111) prefers wetter
habitats, is smaller, and usually has unmarked, darker
hindwings.

Resident

| Jan. | Feb. | Mar. | Apr. | May | June | July | Aug. | Sept. | Oct. | Nov. | Dec. |

**male**

**Dorsal (above)**
orange scaling
yellow orange spots
long straight stigma
faint orange scaling

**Ventral (below)**
dull orange scaling
faint, straight band of
small pale spots

**Ventral**

**Female**

**Larva**

**Comments:** Common throughout most of its range, Peck's Skipper is a butterfly of open, often moist grassy areas where it maneuvers close to the ground with a rapid, darting flight. It tolerates a wide range of human-disturbed, grassy areas including roadsides and suburban lawns. Its distinctive light ventral hindwing patch is somewhat variable in appearance and may be solid or broken into spot bands.

# Peck's Skipper
*Polites peckius*

**Family/Subfamily:** Skippers (Hesperiidae)/ Banded Skippers (Hesperiinae)

**Wingspan:** 1.00–1.25" (2.5–3.2 cm)

**Above:** dark brown; forewing has orange scaling toward base and a few tiny orange spots near apex; hindwing has a band of elongated narrow orange spots

**Below:** variable; hindwing is dark brown with a distinctive irregular central golden yellow patch

**Sexes:** similar, although female has reduced orange scaling at forewing base

**Egg:** whitish green, laid singly on host leaves

**Larva:** dark maroon brown with short light hairs and a black head and anal patch

**Larval Host Plants:** various grasses including Rice Cutgrass and Kentucky Bluegrass

**Habitat:** open grassy areas including pastures, roadsides, old fields, wet meadows, marshes, utility easements and lawns

**Broods:** two or three generations

**Abundance:** occasional to common

**Compare:** Sachem (pg. 293) is larger and has long, narrow wings; male has large rounded black stigma on the forewing above.

Resident

Jan. Feb. Mar. Apr. May June July Aug. Sept. Oct. Nov. Dec.

**Dorsal (above)**
orange scaling
dark brown borders

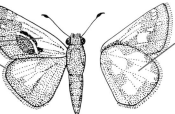

**Ventral (below)**
variable central golden yellow patch; often fused or separated into component spots

**133**

Female

Male pg. 281

**Comments:** The Whirlabout is a diminutive skipper with two distinct rows of squarish spots on the hindwings below. This species is sexually dimorphic. Males are tawny orange and considerably brighter than their drab brown female counterparts. It regularly expands its range each summer, establishing temporary breeding colonies throughout the southeast from Maryland to Texas. Living up to its name, adults have a low, erratic flight and scurry quickly around, periodically stopping to perch or nectar. Avidly fond of flowers, the butterfly is a frequent garden visitor.

# Whirlabout
*Polites vibex*

**Family/Subfamily:** Skippers (Hesperiidae)/ Banded Skippers (Hesperiinae)

**Wingspan:** 1.00–1.25" (2.5–3.2 cm)

**Above:** elongated wings; golden orange with black borders and black stigma; female is dark brown with cream spots on forewing

**Below:** hindwing yellow in male or bronze brown in female with two loose bands of large dark brown spots

**Sexes:** dissimilar; female brown above with little orange scaling; olive brown below with similar pattern as male

**Egg:** white, laid singly on host leaves

**Larva:** brownish green with thin, dark dorsal stripe and black head

**Larval Host Plants:** various grasses including Bermuda Grass, St. Augustine Grass and Crabgrass

**Habitat:** open, disturbed areas including old fields, roadsides, vacant lots, open woodlands, forest edges, parks, lawns and gardens

**Broods:** multiple generations

**Abundance:** occasional to abundant

**Compare:** Fiery Skipper (pg. 279) has more elongated forewings and scattered, tiny dark spots on ventral hindwing.

Resident Visitor

| Jan. | Feb. | Mar. | Apr. | May | June | July | Aug. | Sept. | Oct. | Nov. | Dec. |

**male**

**Dorsal (above)**
golden orange
large black stigma
jagged black border
orange
smooth black border

**Ventral (below)**
two bands of large dark brown spots
golden orange

**135**

subspecies *ontario*

*favonius* Larva

**Comments:** Populations of Southern Hairstreaks outside of the Florida peninsula and extreme southern Atlantic coast to southeastern South Carolina were previously treated as an entirely separate species (the Northern Hairstreak). They are now recognized as geographic races of the same butterfly. Adults have a quick, erratic flight and frequently perch high on the leaves of surrounding vegetation. They regularly venture down to nectar at a variety of small-flowered blossoms.

# Southern Hairstreak
*Fixsenia favonius ontario and F. favonius favonius*

**Family/Subfamily:** Gossamer Wings (Lycaenidae)/
Hairstreaks (Theclinae)

**Wingspan:** 1.0–1.3" (2.5–3.3 cm)

**Above:** dark brown with dark forewing stigma and a
small orange spot near the tails

**Below:** subspecies *ontario* is gray brown with a white
postmedian line that zigzags toward the tails, a large
blue patch often capped lightly in orange, and a short
row on small orange spots; subspecies *favonius* has a
much larger, broad reddish orange patch on the hindwing, a small white spot along the hindwing leading
margin and longer tails

**Sexes:** similar

**Egg:** pinkish brown, laid singly on twigs; eggs overwinter

**Larva:** pale green; covered with tiny yellow dots

**Larval Host Plants:** various oaks

**Habitat:** woodland edges, oak scrub, adjacent open areas

**Broods:** single generation

**Abundance:** occasional to common

**Compare:** White M Hairstreak (pg. 95) has single red
ventral hindwing spot. Gray Hairstreak (pg. 261) is light
gray below with a less jagged white line.

Resident

Jan. Feb. Mar. Apr. May June July Aug. Sept. Oct. Nov. Dec.

male

**Dorsal (above)**
dull, gray brown wings

**Ventral (below)**
jagged black-and-white
line

blue patch with orange
cap

**Comments:** The Striped Hairstreak's name comes from the numerous dark dashes edged in white that give it an overall striped appearance. It is generally uncommon and highly localized. Adults are often encountered with a variety of other early season hairstreaks at available flowers. Females lay the small, flattened eggs singly on host twigs. The eggs overwinter and the young larvae hatch the following spring and feed on the buds and young leaves.

# Striped Hairstreak
*Satyrium liparops*

**Family/Subfamily:** Gossamer Wings (Lycaenidae)/ Hairstreaks (Theclinae)

**Wingspan:** 1.0–1.3" (2.5–3.3 cm)

**Above:** unmarked dark brown with two hindwing tails

**Below:** brown to slate gray with numerous wide, dark bands outlined in white; hindwing has an orange-capped blue patch and several red spots near tails

**Sexes:** similar

**Egg:** pinkish brown, flattened, laid singly on twigs of host

**Larva:** bright green with yellow-green oblique stripes and dark dorsal line

**Larval Host Plants:** various trees and shrubs in the heath and rose families including flame azalea, Highbush Blueberry, Sparkleberry, Black Cherry, Wild Cherry, serviceberry and hawthorn

**Habitat:** mixed deciduous forests, thickets, forest clearings, woodland edges and adjacent open areas

**Broods:** single generation

**Abundance:** occasional

**Compare:** Banded Hairstreak (pg. 129), Hickory Hairstreak (pg. 125) and King's Hairstreak (pg. 127) have less extensive, narrower ventral bands. Banded and Hickory lack the extensive orange cap on the blue hindwing patch.

Resident

Jan. Feb. Mar. Apr. May June July Aug. Sept. Oct. Nov. Dec.

**Dorsal (above)**
unmarked brown

**Ventral (below)**
wide, dark bands outlined in white

blue patch with orange cap

**139**

Ventral    Larva

**Comments:** As its name implies, this is small drab
species is the most common roadside-skipper in the
Carolinas. Nonetheless, this denizen of grassy woodland clearings, forest margins and stream corridors is
rather localized in occurrence and seldom encountered
in large numbers. The adults are often found on low
sunlit perches or at nearby flowers, where they tend
to be rather wary and difficult to closely approach.
Males regularly visit moist earth to imbibe water and
nutrients. Larvae construct individual leaf shelters on
the host. Larvae overwinter.

# Common Roadside-Skipper
*Amblyscirtes vialis*

**Family/Subfamily:** Skippers (Hesperiidae)/
Banded Skippers (Hesperiinae)

**Wingspan:** 1.0–1.3" (2.5–3.3 cm)

**Above:** primarily dark blackish brown with a few small
white spots near the forewing apex

**Below:** dark grayish black with faint gray frosting;
forewing has a few small white spots (sometimes
fused into a tapering band) near the apex

**Sexes:** similar

**Egg:** pale green, laid singly on host leaves

**Larva:** pale green with a whitish head marked with reddish brown vertical lines

**Larval Host Plants:** various grasses including Indian
Woodoats, Bermuda Grass and bentgrass

**Habitat:** open grassy areas in or near woodlands and
along stream corridors

**Broods:** two generations

**Abundance:** occasional

**Compare:** Bell's Roadside-Skipper (pg. 159) has extensive white spotting on the hindwing below.

Resident

Jan. Feb. Mar. Apr. May June July Aug. Sept. Oct. Nov. Dec.

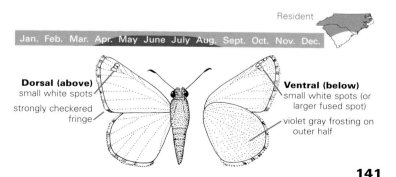

**Dorsal (above)**
small white spots
strongly checkered
fringe

**Ventral (below)**
small white spots (or
larger fused spot)
violet gray frosting on
outer half

**Female**

**Larva**

**Comments:** Hayhurst's Scallopwing is named for the distinctive scalloped hindwing margins that quickly distinguish it from all other dark skippers in the Carolinas. A butterfly of open, disturbed sites including suburban gardens, it is often found in close association with its introduced weedy larval host. The butterfly likely utilizes several native members of the Amaranth family as well. Adults have a quick, scurrying flight and regularly perch or nectar with their wings outstretched, providing a prominent view of their namesake wing shape. Larvae construct individual rolled leaf shelters on the host.

# Hayhurst's Scallopwing
*Staphylus hayhurstii*

**Family/Subfamily:** Skippers (Hesperiidae)/
Spread-wing Skippers (Pyrginae)

**Wingspan:** 1.0–1.4" (2.5–3.6 cm)

**Above:** dark blackish brown with black bands and subtle
gold flecking; forewing has a few small white spots;
hindwing has distinctly scalloped margin

**Below:** forewing is marked as above but paler

**Sexes:** similar, although female is lighter brown with an
overall more banded appearance

**Egg:** pink, laid singly on the underside of host leaves

**Larva:** green with a pinkish hue, a dark green dorsal
stripe, a thin yellow lateral stripe, and a black head

**Larval Host Plants:** Lamb's Quarters

**Habitat:** sunlit wooded stream margins, weedy disturbed sites, fallow agricultural land, forest edges and
gardens

**Broods:** two or more generations

**Abundance:** rare to uncommon

**Compare:** unique

Resident

| Jan. | Feb. | Mar. | Apr. | May | June | July | Aug. | Sept. | Oct. | Nov. | Dec. |

female

Dorsal (above)                                          Ventral (below)
small glassy spots
subtle black bands
checkered fringe
scalloped margin

**143**

**Comments:** The Salt Marsh Skipper's light wing veins and distinct white dash on the hindwing quickly separate it from all other similar-looking skippers. As its name suggests, it is a locally common inhabitant of coastal salt marshes. Adults have a low, rapid flight and frequently visit available flowers along nearby roads or in adjacent fields. Little detailed information is known about the biology of the immature stages.

# Salt Marsh Skipper
*Panoquina panoquin*

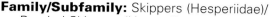

**Family/Subfamily:** Skippers (Hesperiidae)/
Banded Skippers (Hesperiinae)

**Wingspan:** 1.1–1.3" (2.8–3.3 cm)

**Above:** brown with small light spots across forewing

**Below:** light brown with light veins and distinct cream
dash in center of hindwing

**Sexes:** similar

**Egg:** light green, laid singly on host leaves

**Larva:** yellow green

**Larval Host Plants:** grasses including Salt Grass

**Habitat:** salt marshes and adjacent open areas, roadsides

**Broods:** multiple generations

**Abundance:** occasional to common

**Compare:** Brazilian Skipper (pg. 229) and Ocola Skipper
(pg. 213) lack pale ventral wing venation and distinct
cream dash in center of hindwing.

Resident

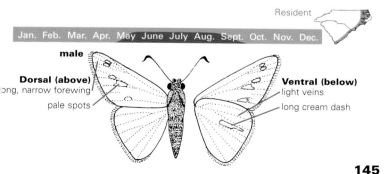

Jan. Feb. Mar. Apr. May June July Aug. Sept. Oct. Nov. Dec.

**male**

**Dorsal (above)**
ong, narrow forewing
pale spots

**Ventral (below)**
light veins
long cream dash

Female

Male pg. 289    Female    Male

**Comments:** Sexually dimorphic, the Zabulon Skipper is a small butterfly with a bright orange male and a drab brown female. Primarily a butterfly of woodlands and adjacent open sites, it occasionally wanders into suburban gardens. Adults have a rapid flight and are strongly attracted to flowers. Males perch on sunlit branches along trails or clearings and engage passing objects or rival males. Females generally prefer to remain within the confines of nearby shady sites.

# Zabulon Skipper
*Poanes zabulon*

**Family/Subfamily:** Skippers (Hesperiidae)/
Banded Skippers (Hesperiinae)

**Wingspan:** 1.0–1.4" (2.5–3.6 cm)

**Above:** male is golden orange with dark brown borders
and small brown spot near forewing apex; female is
dark brown with band of cream spots across forewing

**Below:** male hindwing yellow with a brown base enclosing a yellow spot; female is dark brown with small
light subapical spots, lavender scaling on wing margins, and white bar along leading margin of hindwing

**Sexes:** dissimilar, female brown with little orange color

**Egg:** pale green, laid singly on host leaves

**Larva:** tan with dark dorsal stripe, white lateral stripe and
short, light-colored hairs; reddish brown head

**Larval Host Plants:** various grasses including
Purpletop Grass, Whitegrass and lovegrass

**Habitat:** open woodlands, forest edges, roadsides, pastures, wetland edges, stream corridors and old fields

**Broods:** two generations

**Abundance:** occasional

**Compare:** Female "Pocahontas" form of Hobomok
Skipper (pg. 195) lacks white bar along the leading
margin of the ventral hindwing.

Resident

Jan. Feb. Mar. Apr. May June July Aug. Sept. Oct. Nov. Dec.

male

**Dorsal (above)**
dark spot

narrow black cell
end bar

golden orange

clear golden
orange

**Ventral (below)**
dark base encloses
yellow spot

yellow with darker
spots

**147**

Ventral

Larva

**Comments:** This widespread eastern skipper was once considered the same species as the Southern Broken-Dash. Like its close relative with which it often flies, the species' unique name comes from the dark forewing stigma that is separated or "broken" into two distinct dashes. Encountered in open habitats bordering woodlands, it readily ventures into nearby gardens in search of nectar. Adults are highly active butterflies but frequently stop to perch or feed where they can be easily and closely observed. Larvae overwinter.

# Northern Broken-Dash
*Wallengrenia egeremet*

**Family/Subfamily:** Skippers (Hesperiidae)/
Banded Skippers (Hesperiinae)

**Wingspan:** 1.0–1.5" (2.5–3.8 cm)

**Above:** dark brown; forewing has yellow orange scaling
along the costal margin, a distinct separated stigma
and an elongated bright yellow orange spot extending
outward from the tip of the stigma

**Below:** hindwing is yellow brown with central band of
faint, light spots

**Sexes:** dissimilar; female is primarily dark brown above
with a few elongated cream yellow forewing spots
and reduced yellow orange markings

**Egg:** green, laid singly on host leaves

**Larva:** light green with darker green mottling, yellow lateral stripes; dark brown head with faint vertical stripes

**Larval Host Plants:** various grasses including
Switchgrass and Deertongue

**Habitat:** open, sunny areas near woodlands including
forest clearings and margins, roadsides, fallow agricultural land, pastures, old fields and gardens

**Broods:** one to two generations

**Abundance:** occasional to common

**Compare:** Southern Broken-Dash (pg. 151) is reddish
brown below.

Resident

Jan. Feb. Mar. Apr. May June July Aug. Sept. Oct. Nov. Dec.

male

**Dorsal (above)**
yellow orange spot
at end of stigma

"broken" black
stigma

**Ventral (below)**
dull yellow brown

light spot band

**149**

**Comments:** The Southern Broken-Dash's unusual name comes from the noticeable separation of the forewing stigma that resembles two disconnected black lines. It can be quickly identified by its warm reddish brown ventral hindwings. It is a frequent garden visitor. Fond of flowers, the adults sit quietly on available blossoms and are easy to closely observe or photograph.

# Southern Broken-Dash
*Wallengrenia otho*

**Family/Subfamily:** Skippers (Hesperiidae)/
Banded Skippers (Hesperiinae)

**Wingspan:** 1.0–1.5" (2.5–3.8 cm)

**Above:** male is dark brown with golden orange scaling
along costal margin and in center of hindwing;
forewing has broken black stigma near elongated
orange spot; female is dark brown with several small
cream-orange spots across forewing

**Below:** hindwing reddish brown with faint band of light
spots

**Sexes:** dissimilar; female darker with reduced orange
markings and pale forewing spots

**Egg:** pale green, laid singly on host leaves

**Larva:** light green with dark mottling and dark head

**Larval Host Plants:** various grasses including St.
Augustine Grass and Crabgrass

**Habitat:** moist woodlands, forest edges, wetlands, roadsides, pastures, old fields and gardens

**Broods:** multiple generations

**Abundance:** common to abundant

**Compare:** Sachem (pg. 155) has more elongated wings
and broad pale ventral hindwing patch. Northern
Broken-Dash (pg. 149) is duller brown below.

Resident Visitor

Jan. Feb. Mar. Apr. May June July Aug. Sept. Oct. Nov. Dec.

**male**

**Dorsal (above)**
orange dash

eparated black stigma

**Ventral (below)**
contrasting gray fringe

reddish brown

pale spot band

**151**

**Comments:** The Little Glassywing is named for the translucent, "glassy" spots on the forewing. It regularly explores areas adjacent to its preferred moist woodland habitat. Adults have a low, quick flight and readily visit available flowers. It is not commonly encountered in home gardens. Males perch on low growing vegetation in sunny areas for females.

# Little Glassywing
*Pompeius verna*

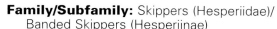

**Family/Subfamily:** Skippers (Hesperiidae)/
Banded Skippers (Hesperiinae)

**Wingspan:** 1.0–1.5" (2.5–3.8 cm)

**Above:** male is dark brown with black stigma and several semitransparent spots across forewing; female is dark brown with several semitransparent spots across forewing

**Below:** dark brown; hindwing purplish brown with band of faint, light spots

**Sexes:** similar; female darker with more rounded wings and larger forewing spots

**Egg:** white, laid singly on host leaves

**Larva:** green to greenish brown with dark mottling and stripes; head is reddish brown

**Larval Host Plants:** grasses including Purpletop Grass

**Habitat:** moist, open woodlands, forest edges, wetlands, roadsides, pastures, old fields and gardens

**Broods:** multiple generations

**Abundance:** occasional to common

**Compare:** Dun Skipper (pg. 167) lacks semitransparent dorsal forewing spots and faint ventral hindwing band.

Resident

Jan. Feb. Mar. Apr. May June July Aug. Sept. Oct. Nov. Dec.

**male**

**Dorsal (above)**
semitransparent spots
(central spot
somewhat square)

black stigma

**Ventral (below)**
purplish brown

faint pale band

Female

Male pg. 293    Female    Male

**Comments:** The Sachem shares its affinity for open, disturbed sites with the Whirlabout and Fiery Skipper
with which it often flies. Adults have a quick, darting
flight that is usually low to the ground. All three
species are somewhat nervous butterflies, but readily
congregate at available flowers. Together, they often
form a circus of activity with several pausing briefly to
perch or nectar before one flies up and disturbs the
others, only to alight again moments later.

# Sachem
*Atalopedes campestris*

**Family/Subfamily:** Skippers (Hesperiidae)/ Banded Skippers (Hesperiinae)

**Wingspan:** 1.0–1.5" (2.5–3.8 cm)

**Above:** elongated wings; male is golden orange with brown borders and large, black stigma; female is dark brown with golden markings in wing centers; forewing has black median spot and several semitransparent spots

**Below:** variable; hindwing golden brown in male, brown in female with pale postmedian patch or band of spots

**Sexes:** dissimilar; female darker with semitransparent forewing spots and reduced orange markings

**Egg:** white, laid singly on host leaves

**Larva:** greenish brown with thin, dark dorsal stripe and black head

**Larval Host Plants:** various grasses including Bermuda Grass and Crabgrass

**Habitat:** open, disturbed areas including old fields, pastures, roadsides, parks, lawns and gardens

**Broods:** multiple generations

**Abundance:** common to abundant

**Compare:** Fiery Skipper (pg. 279) and Whirlabout (pg. 135) have dark spots on the ventral hindwing.

Resident Visitor

| Jan. | Feb. | Mar. | Apr. | May | June | July | Aug. | Sept. | Oct. | Nov. | Dec. |

male

**Dorsal (above)**
elongated forewing
large squarish stigma
golden orange

**Ventral (below)**
large pale patch or postmedian spot band

**Comments:** The small Carolina Satyr is one of the most abundant satyrs in the Southeast. Although it frequents shady woodlands, it regularly strays into nearby residential gardens. Adults have a low, erratic flight and bob slowly among understory vegetation and tall grass. Between periodic bursts of activity, adults perch on grasses or leaf litter with their wings tightly closed and can be easily approached for observation.

# Carolina Satyr
*Hermeuptychia sosybius*

**Family/Subfamily:** Brush-foots (Nymphalidae)/ Satyrs and Wood Nymphs (Satyrinae)

**Wingspan:** 1.0–1.5" (2.5–3.8 cm)

**Above:** dark brown with no distinct markings or eye-spots

**Below:** brown with two narrow, dark brown lines through center of wings; hindwing has row of yellow-rimmed dark eyespots

**Sexes:** similar

**Egg:** green, laid singly on host leaves

**Larva:** pale green with darker green longitudinal stripes and short yellow hairs

**Larval Host Plants:** various grasses including St. Augustine Grass, Broadleaf Carpetgrass and Bermuda Grass

**Habitat:** woodlands, forest clearings, trails and roadsides and adjacent disturbed, grassy areas

**Broods:** multiple generations

**Abundance:** occasional to abundant

**Compare:** Viola's Wood Satyr (pg. 227) and Little Wood Satyr (pg. 215) are larger and have two large, yellow-rimmed dorsal and ventral eyespots on both wings. Gemmed Satyr (pg. 187) lacks ventral eyespots.

Resident

| Jan. | Feb. | Mar. | Apr. | May | June | July | Aug. | Sept. | Oct. | Nov. | Dec. |

**Dorsal (above)**
brown without eyespots

**Ventral (below)**
dark lines
many small eyespots

**157**

Ventral

**Comments:** This small dark skipper has an extremely
limited range in our area, being restricted to the
extreme western tip of South Carolina. It is typically
encountered singly or in small numbers. Males perch
on low vegetation in sunlit sites and readily puddle at
moist ground. The developing larvae construct individual shelters on the host by rolling individual leaves or
tying several leaves together with silk. Fourth instar
larvae from the last generation overwinter and complete development the following spring.

# Bell's Roadside-Skipper
*Amblyscirtes belli*

**Family/Subfamily:** Skippers (Hesperiidae)/
Banded Skippers (Hesperiinae)

**Wingspan:** 1.1–1.4" (2.8–3.6 cm)

**Above:** dark blackish brown with small white forewing
spots

**Below:** dark grayish black with a dusting of white scales,
a white postmedian spot band on the hindwing, and a
checkered fringe

**Sexes:** similar

**Egg:** white, laid singly on host leaves

**Larva:** pale green with a darker green dorsal line and a
whitish head marked with orange brown vertical lines

**Larval Host Plants:** various grasses including St.
Augustine Grass and Indian Woodoats

**Habitat:** open grassy areas in or near woodlands and
along stream corridors

**Broods:** two generations

**Abundance:** rare to occasional

**Compare:** Common Roadside-Skipper (pg. 141) lacks
extensive white spotting on the hindwing below.

Resident

| Jan. | Feb. | Mar. | Apr. | May | June | July | Aug. | Sept. | Oct. | Nov. | Dec. |

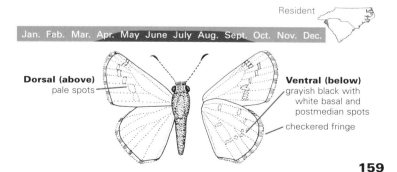

**Dorsal (above)**
pale spots

**Ventral (below)**
grayish black with
white basal and
postmedian spots

checkered fringe

**159**

Dorsal

Larva

**Comments:** Named for the region from which it was first described, the Carolina Roadside-Skipper is a butterfly of dense wet woods, forested swamps and stream corridors that support stands of cane. Generally uncommon throughout its limited southeastern range, it is seldom encountered in large numbers. Colonies tend to be somewhat spotty and localized. Adults are often encountered in sunlit clearings or margins where they tend to alight on vegetation. This remains a poorly known butterfly.

# Carolina Roadside-Skipper
*Amblyscirtes carolina*

**Family/Subfamily:** Skippers (Hesperiidae)/
Banded Skippers (Hesperiinae)

**Wingspan:** 1.1–1.4" (2.8–3.6 cm)

**Above:** dark brown with small cream yellow spots

**Below:** dirty yellow brown with pale forewing spots and
small reddish brown hindwing spots; abdomen has a
row of dark spots

**Sexes:** similar

**Egg:** currently undocumented

**Larva:** gray green with a dark dorsal stripe; reddish head
marked with vertical cream lines

**Larval Host Plants:** Switchcane

**Habitat:** moist woodlands, stream corridors and swamps

**Broods:** three generations

**Abundance:** rare to occasional

**Compare:** Reversed Roadside-Skipper (pg. 169) has a
reversed color pattern with a reddish brown ventral
hindwing with yellow spots.

Resident

Jan. Feb. Mar. Apr. May June July Aug. Sept. Oct. Nov. Dec.

**Dorsal (above)**
small cream yellow
spots

**Ventral (below)**
yellow brown
hindwing

reddish brown spots

Larva

**Comments:** The Clouded Skipper is our only member of
this primarily tropical genus. Although superficially
appearing drab, closer inspection is needed to appreci-
ate the delicate lavender scaling on the hindwings
below. It regularly explores the surrounding landscape
and may show up occasionally in suburban yards or
gardens. Adults have a quick, nervous flight but readily
alight on low vegetation or leaf litter. It is an easy butterfly to approach.

# Clouded Skipper
*Lerema accius*

**Family/Subfamily:** Skippers (Hesperiidae)/ Banded Skippers (Hesperiinae)

**Wingspan:** 1.0–1.5" (2.5–3.8 cm)

**Above:** dark chocolate brown; forewing has several small semitransparent spots

**Below:** dark brown; forewing has row of small semi-transparent spots near apex; both wings have lavender scaling in center and toward margin

**Sexes:** similar, female has larger forewing spots

**Egg:** laid singly on host leaves

**Larva:** green with a dark dorsal stripe and white lateral stripes; head is white with a black margin and three black stripes; body is covered with fine white spots

**Larval Host Plants:** various grasses including St. Augustine Grass, Plumegrass and Rustyseed Paspalum

**Habitat:** open woodlands, forest edges, roadsides, wetland edges, swamps, stream corridors and old fields

**Broods:** multiple generations

**Abundance:** occasional

**Compare:** Zabulon Skipper (pg. 147) female has larger forewing spots and a white bar along leading edge of ventral hindwing. Dusted Skipper (pg. 201) has small white spot near base of ventral hindwing.

Resident Visitor

| Jan. | Feb. | Mar. | Apr. | May | June | July | Aug. | Sept. | Oct. | Nov. | Dec. |

**male**

**Dorsal (above)**
small semitransparent spots

**Ventral (below)**
three small semitransparent spots

lavender frosting

dark central band

Male

Larva

**Comments:** The Cobweb Skipper is named for its distinctly jagged pattern on the ventral hindwing that resembles a spider's web. On the wing for a single spring flight, look for it in clearings, disturbed areas or other grassy sites in close association with stands of its larval hosts. Colonies are typically of low density, spotty and highly localized, although it can be fairly common when encountered. The adults have a rapid and extremely low flight. It is a wary butterfly but readily perches on or near the ground and visits a variety of low-growing spring flowers. Larvae construct individual leaf shelters on the host. Mature larvae overwinter.

# Cobweb Skipper
*Hesperia metea*

**Family/Subfamily:** Skippers (Hesperiidae)/
Banded Skippers (Hesperiinae)

**Wingspan:** 1.1–1.4" (2.8–3.6 cm)

**Above:** olive brown; forewing has tawny orange spots
and a narrow black stigma; hindwing has angled band
of tawny orange spots

**Below:** hindwing is olive brown with irregular white
bands and veins giving an overall cobweb appearance

**Sexes:** dissimilar; female is primarily dark brown with a
few pale forewing spots near the apex

**Egg:** white, laid singly on or near host leaves

**Larva:** gray brown with a dark dorsal stripe and a round
black head

**Larval Host Plants:** various grasses including Little
Bluestem and Big Bluestem

**Habitat:** open areas within woodlands, pine barrens, utility easements and recently cleared or burned sites

**Broods:** single generation

**Abundance:** rare to occasional

**Compare:** unique

Resident

Jan. Feb. Mar. Apr. May June July Aug. Sept. Oct. Nov. Dec.

**male**

**Dorsal (above)**
narrow black stigma

orange scaling

angled band of tawny
orange spots

**Ventral (below)**
"cobweb"
appearance—
irregular white
bands and veins

**165**

Male

Larva

**Comments:** The Dun Skipper is a small chocolate brown butterfly with few markings. Although it prefers moist, grassy or sedge-dominated areas associated with deciduous woods, it frequently ventures into surrounding areas, and is periodically encountered in home gardens. Adults have a quick, low flight and dart around erratically over the vegetation. Males occasionally visit damp ground.

# Dun Skipper
*Euphyes vestris*

**Family/Subfamily:** Skippers (Hesperiidae)/ Banded Skippers (Hesperiinae)

**Wingspan:** 1.0–1.5" (2.5–3.8 cm)

**Above:** male is dark chocolate brown with black stigma; female is dark brown with several whitish spots on forewing

**Below:** brown; typically unmarked hindwing, but occasionally has faint spot band

**Sexes:** similar; female has small, white forewing spots

**Egg:** green, laid singly on host leaves

**Larva:** green with thin white lines; head is brown with light outer stripes and dark center

**Larval Host Plants:** various sedges including Upright Sedge

**Habitat:** moist areas in or near deciduous woodlands, wet meadows, marshes, forest edges, stream corridors, roadsides, pastures, utility easements and old fields

**Broods:** two generations

**Abundance:** occasional

**Compare:** Little Glassywing (pg. 153) has distinct glassy white spots on forewing and defined ventral hindwing band.

Resident

Jan. Feb. Mar. Apr. May June July Aug. Sept. Oct. Nov. Dec.

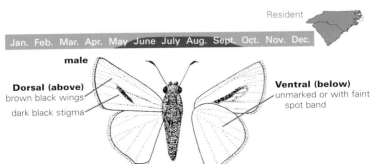

male

**Dorsal (above)**
brown black wings

dark black stigma

**Ventral (below)**
unmarked or with faint spot band

Dorsal

Larva

**Comments:** Most of this skipper's limited southeastern range occurs within the Carolinas. With reddish brown hindwings and yellow spots, it has the reversed pattern of the Carolina Roadside-Skipper and was once considered to be a mere color form of that species. It, too, is a butterfly of wet woodlands, forested swamps and stream corridors that support abundant stands of cane. Little detailed information is known about this rare butterfly's biology and behavior. Although it has been reared, the details of its life history remain poorly known.

# Reversed Roadside-Skipper
*Amblyscirtes reversa*

**Family/Subfamily:** Skippers (Hesperiidae)/
Banded Skippers (Hesperiinae)

**Wingspan:** 1.1–1.4" (2.8–3.6 cm)

**Above:** dark brown with small cream yellow spots

**Below:** reddish brown with pale yellow spots

**Sexes:** similar

**Egg:** currently undocumented

**Larva:** gray green with dark dorsal stripe; reddish head
marked with pale vertical stripes

**Larval Host Plants:** Switchcane

**Habitat:** moist woodlands, stream corridors and swamps

**Broods:** two generations

**Abundance:** rare to occasional

**Compare:** Carolina Roadside-Skipper (pg. 161) has a yellow ventral hindwing with small reddish brown spots.

Resident

Jan. Feb. Mar. Apr. May June July Aug. Sept. Oct. Nov. Dec.

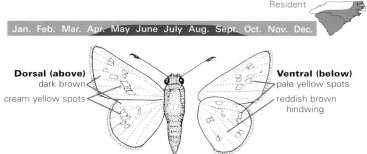

**Dorsal (above)**
dark brown
cream yellow spots

**Ventral (below)**
pale yellow spots
reddish brown
hindwing

**169**

Male

Larva

**Comments:** The Dreamy Duskywing is a small, early
season species that just barely makes its way into our
fauna in the western Carolinas. The Dreamy and the
similar but larger and more widespread Sleepy
Duskywing, lack glassy forewing spots that character-
ize all other members of this group. Adults scurry low
to the ground along forest trails or margins with a fast,
bouncing flight. Males are frequently encountered puddling at moist areas. Larvae construct individual leaf
shelters on host and overwinter in a similar structure.

# Dreamy Duskywing
*Erynnis icelus*

**Family/Subfamily:** Skippers (Hesperiidae)/
Spread-wing Skippers (Pyrginae)

**Wingspan:** 1.0–1.6" (2.5–4.1 cm)

**Above:** dark brown; forewing has extensive gray scaling
toward outer margin, two black chain-like bands
enclosing a broad gray patch, a dark base, lacks glassy
spots; hindwing has two rows of pale spots; labial
palpi are noticeably long and project forward

**Below:** dark brown, hindwing has two rows of pale spots

**Sexes:** similar, although female is lighter with more heavily patterned forewings

**Egg:** green turning reddish, laid singly on stems or leaves

**Larva:** pale green with a dark dorsal stripe, a white lateral stripe, numerous tiny white tubercles, and a black head marked with yellow and red spots

**Larval Host Plants:** various willows, poplars and
birches

**Habitat:** open woodlands, forest edges and clearings,
roadsides, moist woodland depressions

**Broods:** single generation

**Abundance:** uncommon to occasional

**Compare:** Sleepy Duskywing (pg. 197) flies earlier and is
larger with shorter, more rounded forewings.

Resident

| Jan. | Feb. | Mar. | Apr. | May | June | July | Aug. | Sept. | Oct. | Nov. | Dec. |

male

long labial palpi

**Dorsal (above)**
forewing lacks glassy
spots

two black bands

two rows of pale
spots

**Ventral (below)**
two rows of pale
spots

**171**

**Comments:** The Twin-Spot Skipper's unique pattern quickly distinguishes it from all other skippers in the state. A large skipper of coastal marshes and pine forests, it occasionally shows up in home gardens. Adults have a quick, darting flight but regularly alight on low vegetation. Little detailed information is known about the biology and behavior of this locally common butterfly.

# Twin-Spot Skipper
*Oligoria maculata*

**Family/Subfamily:** Skippers (Hesperiidae)/
Banded Skippers (Hesperiinae)

**Wingspan:** 1.25–1.40" (3.2–3.6 cm)

**Above:** dark chocolate brown with small white spots
across forewing

**Below:** brown with three distinct white hindwing spots;
two close together and one isolated

**Sexes:** similar

**Egg:** brownish, laid singly on host leaves

**Larva:** pinkish green with brown head

**Larval Host Plants:** various grasses

**Habitat:** marshes, forest edges and pinelands

**Broods:** multiple generations

**Abundance:** occasional

**Compare:** Clouded Skipper (pg. 163) and Little
Glassywing (pg. 153) are smaller and lack the distinct
ventral hindwing white spots.

Resident

Jan. Feb. Mar. Apr. May June July Aug. Sept. Oct. Nov. Dec.

**Dorsal (above)**
white spots

**Ventral (below)**
brown with three
distinct white
spots: two next
to each other and
one separate

**173**

Male

Ventral  Larva

**Comments:** This primarily northeastern skipper is named for the two small white spots on the upper surface of the forewing in females. The Two-Spotted Skipper is rare and highly localized in occurrence within the Carolinas, being limited to the Atlantic Coastal Plain. A fairly plain butterfly, it may be readily identified by the light veins and bright white anal margin on the hindwing below. Males perch low of sedges or grasses and readily fly out to engage other males. Larvae construct individual leaf shelters on the host. Larvae overwinter.

# Two-Spotted Skipper
*Euphyes bimacula*

**Family/Subfamily:** Skippers (Hesperiidae)/
Banded Skippers (Hesperiinae)

**Wingspan:** 1.25–1.40" (3.2–3.6 cm)

**Above:** dark brown with white fringes; forewing has
small tawny orange patch and black stigma

**Below:** brownish orange with pale veins and a whitish
anal hindwing margin

**Sexes:** dissimilar; female is primarily dark brown with
two cream spots in the center of the forewing above

**Egg:** green, laid singly on host leaves

**Larva:** pale green with a darker dorsal stripe and numerous tiny wavy white dashes; reddish brown head
marked with a black oval ringed in cream on the forehead, and a cream band around the outer margin

**Larval Host Plants:** various sedges including Upright
Sedge

**Habitat:** wet meadows, marshes and swamps near
woodlands

**Broods:** two generations

**Abundance:** rare to occasional

**Compare:** Berry's Skipper (pg. 315) has more orange
scaling above, a restricted range and lacks the white
wing fringes.

Resident

| Jan. | Feb. | Mar. | Apr. | May | June | July | Aug. | Sept. | Oct. | Nov. | Dec. |

male

**Dorsal (above)**
tawny orange scaling
white fringe

**Ventral (below)**
paler veins
white anal margin

Dorsal

Larva

**Comments:** This skipper's white cobweb pattern on the gray brown ventral hindwing is distinctive and difficult to confuse with any other butterfly. It is an inhabitant of moist woodlands and associated margins or road-sides with cane. The species produces two generations and is on the wing in early summer and again toward early fall. As is the case with most road-side-skippers, little detailed information is know about the butterfly's biology and behavior.

# Lace-Winged Roadside-Skipper
*Amblyscirtes aesculapius*

**Family/Subfamily:** Skippers (Hesperiidae)/
Banded Skippers (Hesperiinae)

**Wingspan:** 1.2–1.5" (3.0–3.8 cm)

**Above:** dark brown with checkered fringes and a band of
small white spots on the forewing

**Below:** hindwing is gray brown with checkered fringes
and a network of white bands and veins, giving an
overall lacy appearance

**Sexes:** similar

**Egg:** currently undocumented

**Larva:** blue green with a dark dorsal stripe; reddish
brown head marked with pale crescents on each side

**Larval Host Plants:** Switchcane

**Habitat:** moist dense woodlands

**Broods:** two generations

**Abundance:** rare to common

**Compare:** Cobweb Skipper (pg. 165) lacks checkered
fringes, has less extensive pattern on ventral hindwing
and inhabits dry, open grassy areas.

Resident

| Jan. | Feb. | Mar. | Apr. | May | June | July | Aug. | Sept. | Oct. | Nov. | Dec. |

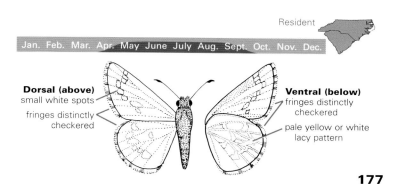

**Dorsal (above)**
small white spots

fringes distinctly
checkered

**Ventral (below)**
fringes distinctly
checkered

pale yellow or white
lacy pattern

**177**

Ventral

Larva

**Comments:** Aptly named, the Confused Cloudywing is often very tricky to accurately distinguish from the more abundant Southern and Northern Cloudywings with which it often flies. Causing much of the problem is its band of glassy forewing spots that varies from highly reduced to prominent. An uncommon and often overlooked southern butterfly, it wanders northward each year to temporarily colonize many locations. Many aspects of the species' life history and behavior remain poorly known.

## Confused Cloudywing
*Thorybes confusis*

**Family/Subfamily:** Skippers (Hesperiidae)/
Spread-wing Skippers (Pyrginae)

**Wingspan:** 1.2–1.6" (3.0–4.1 cm)

**Above:** brown with a variable band of several glassy
white spots across the forewing and a light, checkered
fringe; hindwing is tapered slightly toward bottom;
antennal clubs are all brown

**Below:** forewing is marked as above but paler and with
gray frosting along outer margin; hindwing has two
faint dark brown bands across center and faint light
frosting along outer margin; pale face

**Sexes:** similar

**Egg:** pale green, laid singly on host leaves

**Larva:** greenish brown with a dark dorsal stripe and pale
lateral stripes; blackish head

**Larval Host Plants:** various legumes such as beggarweeds and bush clover

**Habitat:** open sites in or near woodlands including clearings, trails, old fields, pastures and forest margins

**Broods:** two generations

**Abundance:** uncommon to occasional

**Compare:** Northern Cloudywing (pg. 185) has a dark
face. Southern Cloudywing (pg.181) has a prominent,
straight forewing spot band and a white spot on the
antennal club.

Resident Visitor

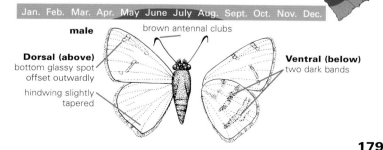

Jan. Feb. Mar. Apr. May June July Aug. Sept. Oct. Nov. Dec.

male

brown antennal clubs

**Dorsal (above)**
bottom glassy spot
offset outwardly

hindwing slightly
tapered

**Ventral (below)**
two dark bands

**179**

Ventral

**Comments:** The Southern Cloudywing frequents forest
edges and open disturbed sites but is also a regular
garden visitor. Adults have a strong, erratic flight but
frequently stop to nectar. Males perch on low vegetation and aggressively dart out to investigate intruders
before returning to the same general location
moments later. Adults rest with their wings partially
open but are often rather nervous and difficult to
closely approach.

# Southern Cloudywing
*Thorybes bathyllus*

**Family/Subfamily:** Skippers (Hesperiidae)/
Spread-wing Skippers (Pyrginae)

**Wingspan:** 1.2–1.6" (3.0–4.1 cm)

**Above:** brown with straight, glassy white spots across
forewing and light, checkered wing fringe; hindwing
tapered slightly toward bottom; antennal club has a
white spot

**Below:** brown; hindwing darker at base with two dark
brown bands; light face

**Sexes:** similar

**Egg:** green, laid singly on the leaves of host

**Larva:** greenish brown with black head, thin, dark dorsal
stripe and narrow light lateral stripe; body covered
with numerous short, light-colored hairs

**Larval Host Plants:** various legumes (Fabaceae) including beggarweeds, bush clovers, Butterfly Pea and Hog
Peanut, Groundnut and milk vetch

**Habitat:** open, disturbed sites including roadsides, old
fields, forest edges, meadows and gardens

**Broods:** multiple generations

**Abundance:** occasional to common

**Compare:** Northern Cloudywing (pg. 185) and Confused
Cloudywing (pg. 179) have smaller white dorsal
forewing spots.

Resident

Jan. Feb. Mar. Apr. May June July Aug. Sept. Oct. Nov. Dec.

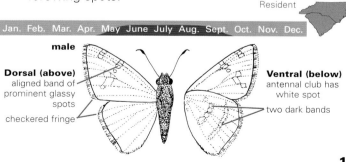

male

**Dorsal (above)**
aligned band of
prominent glassy
spots

checkered fringe

**Ventral (below)**
antennal club has
white spot

two dark bands

**181**

Larva

**Comments:** Finally, an easy-to-identify duskywing! This small skipper can reliably be told apart from all other members of this complicated genus by its distinctively mottled wings. It is a rare to uncommon butterfly throughout its range and typically found in isolated, sporadic colony sites. In the Carolinas, the Mottled Duskywing can be encountered in habitats where the terrain is somewhat hilly and undulating. Males often puddle at damp ground with other duskywings but are rarely the most prevalent species present. Larvae from second brood overwinter in individual leaf shelters.

# Mottled Duskywing
*Erynnis martialis*

**Family/Subfamily:** Skippers (Hesperiidae)/
Spread-wing Skippers (Pyrginae)

**Wingspan:** 1.2–1.6" (3.0–4.1 cm)

**Above:** brown and strongly patterned with numerous
dark blotches giving a distinctive mottled appearance;
forewing has several small glassy spots toward the
apex; fresh individuals have a subtle violet sheen

**Below:** brown with numerous light and dark spots

**Sexes:** similar, although female is lighter with increased
gray scaling and more heavily patterned wings

**Egg:** green soon turning pinkish, laid singly on host
leaves

**Larva:** pale green with numerous tiny white tubercles
and a black head marked with orange spots around the
margin

**Larval Host Plants:** New Jersey Tea

**Habitat:** open upland woodland, forest edges and clearings, barrens and old fields

**Broods:** two generations

**Abundance:** rare to uncommon

**Compare:** unique

Resident

Jan. Feb. Mar. Apr. May June July Aug. Sept. Oct. Nov. Dec.

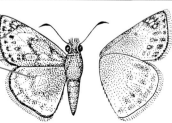

**Dorsal (above)**
glassy spots

both wings with
strong mottling

sometimes has a
faint violet sheen

**Ventral (below)**

**Comments:** The Northern Cloudywing is one of several similar-looking and closely related dark brown skippers found in the Carolinas. The butterfly has a low, skipping flight and quickly darts along trails or clearings occasionally stopping to nectar. It readily visits nearby gardens. Males perch on the ground or low on vegetation and quickly speed out to engage rival males or passing females. At rest, they hold their wings in a relaxed, partially open position.

# Northern Cloudywing
*Thorybes pylades*

**Family/Subfamily:** Skippers (Hesperiidae)/
Spread-wing Skippers (Pyrginae)

**Wingspan:** 1.2–1.7" (3.0–4.3 cm)

**Above:** brown with several small, elongated, misaligned
glassy white spots on forewing and light, checkered
wing fringe

**Below:** brown; hindwing darker at base with two dark
brown bands; forewing has lavender gray scaling
toward apex; dark face

**Sexes:** similar

**Egg:** pale greenish white, laid singly on the leaves of
host

**Larva:** greenish brown with black head, thin, dark dorsal
stripe and narrow pinkish brown lateral stripe; body
covered with numerous short, light-colored hairs

**Larval Host Plants:** various legumes including beggarweeds, bush clovers, milk vetch, alfalfa and clovers

**Habitat:** open woodlands, forest clearings, forest edges,
roadsides, utility easements and old fields

**Broods:** two generations

**Abundance:** occasional to common

**Compare:** Southern Cloudywing (pg. 181) has larger dorsal forewing spots that form an aligned band.

Resident

Jan. Feb. Mar. Apr. May June July Aug. Sept. Oct. Nov. Dec.

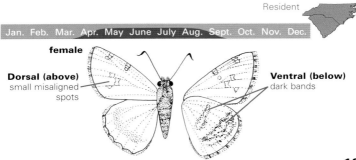

female

**Dorsal (above)**
small misaligned
spots

**Ventral (below)**
dark bands

Larva

**Comments:** Often overlooked, the Gemmed Satyr is
one of our most attractive satyrs and the only one
without eyespots. It dances along the forest floor with
a weak, low flight. Although they can be challenging to
follow through the understory vegetation, adults regularly alight on grass blades or leaf litter. It is seldom
very abundant, occurring in spotty, localized colonies,
and may be overlooked due to its resemblance to the
Carolina Satyr. The slender larvae have both a green
and brown form.

# Gemmed Satyr
*Cyllopsis gemma*

**Family/Subfamily:** Brush-foots (Nymphalidae)/
Satyrs and Wood Nymphs (Satyrinae)

**Wingspan:** 1.25–1.70" (3.2–4.3 cm)

**Above:** warm brown with very small dark spots along
hindwing margin

**Below:** light speckled brown with two narrow, dark wavy
lines through the center of the wings; hindwing has a
large purplish patch containing black spots with silver
highlights

**Sexes:** similar

**Egg:** green, laid singly on host leaves

**Larva:** green or brown with thin pale stripes, two small
tails and two brownish pink horns on the head

**Larval Host Plants:** Bermuda Grass and other grasses

**Habitat:** moist woodlands. stream corridors and associated shady grassy areas

**Broods:** two or more generations

**Abundance:** occasional

**Compare:** Carolina Satyr (pg. 157) has a row of yellow-
rimmed ventral eyespots on both the forewing and
hindwing.

Resident

Jan. Feb. Mar. Apr. May June July Aug. Sept. Oct. Nov. Dec.

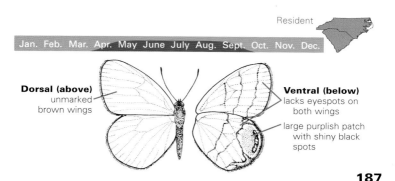

**Dorsal (above)**
unmarked
brown wings

**Ventral (below)**
lacks eyespots on
both wings

large purplish patch
with shiny black
spots

**187**

**Comments:** This abundant skipper commonly wanders into disturbed sites near its habitat in search of nectar. It is often a common garden visitor. It holds its wings outstretched when feeding or at rest. Adults flutter about with quick, somewhat erratic flight usually low to the ground. Males perch on low vegetation and occasionally puddle at damp sand or gravel. The larvae construct individual shelters on the host by folding over leaves with silk. Once fully mature, the larvae from the late season generation overwinter among the leaf litter until the following spring.

# Horace's Duskywing
*Erynnis horatius*

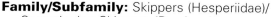

**Family/Subfamily:** Skippers (Hesperiidae)/
Spread-wing Skippers (Pyrginae)

**Wingspan:** 1.25–1.75" (3.2–4.4 cm)

**Above:** dark brown overall with gray scaling usually lacking; forewing has cluster of small clear spots near apex and one at end of forewing cell; female is lighter with more contrasting pattern

**Below:** brown with faint rows of light spots along outer edge of hindwing

**Sexes:** similar, although female is lighter with more heavily patterned forewings and larger forewing spots

**Egg:** pale yellow green, laid singly on new growth of host

**Larva:** pale green with tiny white spots; reddish brown head marked with orange spots around the margin

**Larval Host Plants:** various oaks including Live Oak, Turkey Oak, Water Oak, Myrtle Oak, Scrub Oak and Post Oak

**Habitat:** oak woodlands, oak scrub, forest edges, woodland clearings, roadsides and utility easements

**Broods:** two to three generations

**Abundance:** occasional to common

**Compare:** Juvenal's Duskywing (pg. 217) has two light spots along leading margin of ventral hindwing and extensive gray scaling on dorsal forewing.

Resident

| Jan. | Feb. | Mar. | Apr. | May | June | July | Aug. | Sept. | Oct. | Nov. | Dec. |

**male**

Dorsal (above)
glassy spots

glassy spot at end
of cell

lacks extensive gray
scaling

Ventral (below)
lacks spots along
leading margin

**189**

Male

Female

Larva

**Comments:** The Wild Indigo Duskywing is named for one of its preferred larval hosts. It has become more widespread because of its ability to colonize and use plantings of the introduced groundcover crown vetch. It is often extremely abundant in late summer and early fall. At such time, I have found it easy to locate several larvae and their leaf shelters on virtually every plant of Carolina Indigo examined. Larvae from the last generation overwinter and complete development in the spring. The butterfly has a fast, erratic flight but readily stops to nectar or perch on sunlit leaves or twigs. Males frequently puddle at damp ground.

# Wild Indigo Duskywing
*Erynnis baptisiae*

**Family/Subfamily:** Skippers (Hesperiidae)/ Spread-wing Skippers (Pyrginae)

**Wingspan:** 1.3–1.7 (3.3–4.3 cm)

**Above:** dark brown with pale spots; forewing has darker base with several small glassy spots toward the apex and a distinctive reddish brown patch at the end of the cell; hindwing typically has faint cell-end bar

**Below:** dark brown, hindwing has two rows of prominent light spots

**Sexes:** similar, although female is lighter with more heavily pattered wings and larger glassy spots

**Egg:** green, laid singly on host leaves

**Larva:** pale green with numerous tiny white tubercles, dark dorsal stripe, white lateral stripe; dark brown head marked with pale orange or yellow around the margin

**Larval Host Plants:** wide variety including Carolina Indigo, Wild Indigo, Blue Wild Indigo, White Wild Indigo, Canadian Milkvetch, Wild Lupine and Crown Vetch

**Habitat:** open woodlands, forest edges and clearings, roadsides, utility easements, old fields and roadsides

**Broods:** two or three generations

**Abundance:** occasional to common

**Compare:** Horace's Duskywing (pg. 189) has glassy spot at end of forewing cell.

Resident

Jan. Feb. Mar. Apr. May June July Aug. Sept. Oct. Nov. Dec.

male

**Dorsal (above)**
red-brown spot at end of cell

typically lacks glassy spot at end of cell

basal half of wing darker

faint cell-end bar

**Ventral (below)**
two rows of white spots

191

**Comments:** A butterfly of pine and oak barrens and adjacent meadows, the Dotted Skipper is typically rare and highly localized within the eastern Carolinas. When encountered, this dull olive brown skipper is fairly easy to identify by its distinctive small white dots on the ventral hindwing. Additional studies are needed to unravel the details of this reclusive species' life history, ecology and behavior.

# Dotted Skipper
*Hesperia attalus*

**Family/Subfamily:** Skippers (Hesperiidae)/
Banded Skippers (Hesperiinae)

**Wingspan:** 1.4–1.6" (3.6–4.1 cm)

**Above:** dark brown; forewing is elongated and pointed
with several tawny orange spots; hindwing has postmedian band of small tawny orange spots

**Below:** hindwing is olive brown with a band of small
white dots

**Sexes:** similar, female is primarily dark brown with several pale forewing spots on both wings

**Egg:** laid singly on or near host leaves

**Larva:** green brown with a dark dorsal stripe; black head

**Larval Host Plants:** various grasses including
Arrowfeather Threeawn and Carolina Crabgrass

**Habitat:** sandy pine barrens, woodland meadows

**Broods:** two generations

**Abundance:** rare to occasional

**Compare:** Leonard's Skipper (pg. 211) is larger, reddish
brown beneath with prominent, large white hindwing
spots. Forewings above have tawny orange scaling at
base.

Resident

Jan. Feb. Mar. Apr. May June July Aug. Sept. Oct. Nov. Dec.

**Dorsal (above)**
pointed forewing

**Ventral (below)**
band of tiny white
dots

193

Female "Pocahontas"

Male pg. 305    Female    Ventral    Larva

**Comments:** This small woodland skipper has a single
early summer flight. Males perch on sunlit leaves and
aggressively dart out at other passing butterflies.
Although common throughout the north and east, its
range barely enters the extreme western mountains of
the Carolinas. The female produces two distinct forms.
The lighter form closely resembles the male and the
darker form "Pocahontas" is superficially similar to
Zabulon Skipper females. Larvae construct individual
leaf shelters on the host. Larvae overwinter and complete development the following spring.

# Hobomok Skipper
*Poanes hobomok*

**Family/Subfamily:** Skippers (Hesperiidae)/
Banded Skippers (Hesperiinae)

**Wingspan:** 1.4–1.6" (3.6–4.1 cm)

**Above:** golden orange with irregular dark brown borders
and a narrow black cell-end bar on the forewing

**Below:** purplish brown with a broad yellow orange patch
through the hindwing

**Sexes:** dissimilar; female has two forms. Normal form
resembles male but has reduced orange scaling above.
"Pocahontas" form is dark brown above with pale
forewing spots; hindwing is purplish brown below with
faint band and violet gray frosting along outer margin.

**Egg:** white, laid singly on host leaves

**Larva:** brown green with numerous short, light-colored
hairs; round, brown head

**Larval Host Plants:** various grasses including Little
Bluestem, panic grasses, Poverty Oatgrass, bluegrass
and Rice Cutgrass

**Habitat:** open woodlands, forest edges, clearings and
trails, roadsides and along forested stream margins

**Broods:** single generation

**Abundance:** occasional to common

**Compare:** Female Zabulon Skipper (pg. 147) has a white
bar along leading edge of ventral hindwing.

Resident

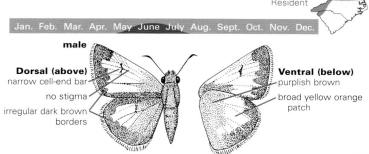

Jan. Feb. Mar. Apr. May June July Aug. Sept. Oct. Nov. Dec.

male

**Dorsal (above)**
narrow cell-end bar
no stigma
irregular dark brown
borders

**Ventral (below)**
purplish brown
broad yellow orange
patch

**195**

Male

Female

Larva

**Comments:** An early season species, it has a single spring flight and is typically out before the smaller and similar-looking Dreamy Duskywing. Together, they are the only two duskywings that lack small glassy forewing spots. It is a widespread butterfly throughout the Carolinas but occasionally goes through fluctuations in population numbers and abundance. Adults perch and feed with their mottled gray brown wings held in an open posture, a common characteristic of all duskywings. Larvae construct individual leaf shelters and overwinter inside.

# Sleepy Duskywing
*Erynnis brizo*

**Family/Subfamily:** Skippers (Hesperiidae)/
Spread-wing Skippers (Pyrginae)

**Wingspan:** 1.30–1.75" (3.3–4.4 cm)

**Above:** dark brown; forewing has extensive gray scaling
toward outer margin and two black chain-like bands;
hindwing has two rows of faint pale spots

**Below:** dark brown, hindwing has two often faint rows
of pale spots

**Sexes:** similar, female is lighter with more heavily patterned forewing and more prominent hindwing spots

**Egg:** green, laid singly on host leaves

**Larva:** pale green with a yellow white lateral stripe and
numerous tiny white tubercles; brown head marked
with six orange spots around the margin

**Larval Host Plants:** various oaks including Scrub Oak
and Black Oak, also American Chestnut

**Habitat:** open oak woodlands and scrub, forest edges
and clearings, roadsides and adjacent open sites

**Broods:** single generation

**Abundance:** uncommon to common

**Compare:** Dreamy Duskywing (pg. 171) is smaller,
forewing has darker base and more distinct gray patch
between black bands; flies later.

Resident

Jan. Feb. Mar. Apr. May June July Aug. Sept. Oct. Nov. Dec.

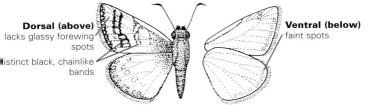

**Dorsal (above)**
lacks glassy forewing
spots

distinct black, chainlike
bands

**Ventral (below)**
faint spots

Female

Male pg. 309

Ventral

Larva

**Comments:** The Yehl Skipper is rare to uncommon large butterfly of southeastern forested swamps or moist woodlands. It is most frequently encountered in sunlit clearings, margins or roadsides. The adults are very wary and often difficult to closely approach. They visit a variety of moisture-loving flowers. Little detailed information is known about the biology and behavior of this reclusive species.

# Yehl Skipper
*Poanes yehl*

**Family/Subfamily:** Skippers (Hesperiidae)/
Banded Skippers (Hesperiinae)

**Wingspan:** 1.30–1.75" (3.3–4.3 cm)

**Above:** golden orange with broad dark brown borders;
forewing has long, narrow black stigma; female is
overall brownish

**Below:** hindwing is yellow orange with a pale band of
spots (usually three prominent spots)

**Sexes:** dissimilar; female has pale forewing spots and
reduced golden orange scaling on the wing bases
above; ventral hindwing is brown with pale spot band

**Egg:** currently undocumented

**Larva:** green brown with a dark dorsal stripe; dark brown
head

**Larval Host Plants:** currently undocumented

**Habitat:** forested swamps and associated clearings and
margins

**Broods:** two generations

**Abundance:** rare to occasional to abundant

**Compare:** Broad-Winged Skipper (pg. 327) has broad
pale ray through the spot band on the hindwing below.

Resident

Jan. Feb. Mar. Apr. May June July Aug. Sept. Oct. Nov. Dec.

**male**

**Dorsal (above)**
long, narrow black
stigma

broad, dark brown
borders

**Ventral (below)**
three to four pale
spots in a band

**199**

Ventral | loammi | loammi ventral | Larva

**Comments:** A butterfly of open, dry habitats, the
Dusted Skipper is uncommon and typically quite localized in occurrence. On the wing in late spring or early
summer, it has but one generation each year. The
elongated slender larvae construct individual shelters
on the host by tying several grass blades together with
silk. The fully grown larvae overwinter in a similar shelter at the base of the host grass clump and pupate the
following spring. Subspecies *loammi*, considered a
separate species by some, occurs locally in coastal
portions of the Carolinas.

## Dusted Skipper
*Atrytonopsis hianna* and *A. hianna loammi*

**Family/Subfamily:** Skippers (Hesperiidae)/
Banded Skippers (Hesperiinae)

**Wingspan:** 1.4–1.7" (3.6–4.3 cm)

**Above:** dark chocolate brown with small glassy forewing
spots

**Below:** dark brown with gray frosting toward outer margin; forewing has small white forewing spots near
apex; hindwing has tiny white spot near base; face is
white with black mask; subspecies *loammi* has angled
bands of white spots

**Sexes:** similar

**Egg:** yellow, laid singly on host leaves

**Larva:** gray, pinkish dorsally, with numerous cream hairs,
a brown anal segment; reddish purple head

**Larval Host Plants:** various grasses including Little
Bluestem and Big Bluestem

**Habitat:** dry habitats including open woodlands, barrens,
utility easements and old fields

**Broods:** single generation

**Abundance:** rare to occasional

**Compare:** Clouded Skipper (pg. 163) lacks small white
spot at base of ventral hindwing. Female Zabulon
Skipper (pg. 147) has white bar along the leading edge
of the ventral hindwing.

Resident

Jan. Feb. Mar. Apr. May June July Aug. Sept. Oct. Nov. Dec.

**Dorsal (above)**
glassy white spots

**Ventral (below)**
dusted gray

single white spot

**201**

**Comments:** Mitchell's Satyr, also called St. Francis' Satyr, is the only federally endangered butterfly in the Carolinas and one of the rarest species in North America. Known only from Fort Bragg, the entire population of the butterfly occurs within only a few square miles. An inhabitant of moist, open meadows with few trees, the adults have a slow, bobbing flight and maneuver low among the wetland vegetation, stopping frequently to perch. Little detailed information is currently is know about the species' biology, behavior or ecology.

# Mitchell's Satyr (St. Francis' Satyr)
*Neonympha mitchellii francisci*

**Family/Subfamily:** Brush-foots (Nymphalidae)/ Satyrs and Wood Nymphs (Satyrinae)

**Wingspan:** 1.3–1.8" (3.3–4.6 cm)

**Above:** uniform unmarked brown

**Below:** brown with a row of round to oval yellow-rimmed black eyespots with silvery blue highlights, and two wavy orange brown lines across both wings

**Sexes:** similar

**Egg:** pearly green, laid singly or in small groups on host leaves

**Larva:** currently undocumented

**Larval Host Plants:** currently undocumented, but likely various sedges

**Habitat:** open wet meadows in sandhills

**Broods:** two generations

**Abundance:** rare

**Compare:** Georgia Satyr (pg. 207) has very elongated (not round) yellow-rimmed black hindwing eyespots and lacks distinct ventral forewing eyespots.

Resident

| Jan. | Feb. | Mar. | Apr. | May | June | July | Aug. | Sept. | Oct. | Nov. | Dec. |

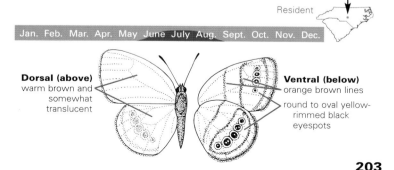

**Dorsal (above)**
warm brown and somewhat translucent

**Ventral (below)**
orange brown lines

round to oval yellow-rimmed black eyespots

Dorsal

**Comments:** The Hoary Edge is named for the large frosty white (or hoary) patch on the ventral hindwing. It readily ventures into disturbed sites near its habitat and shows up occasionally in home gardens. The Hoary Edge is common, but rarely found in any large numbers. Adults have a low, strong flight and can be a challenge to follow. Males perch on low, protruding vegetation and aggressively investigate and chase passing insects.

# Hoary Edge
*Achalarus lyciades*

**Family/Subfamily:** Skippers (Hesperiidae)/
Spread-wing Skippers (Pyrginae)

**Wingspan:** 1.40–1.75" (3.6–4.4 cm)

**Above:** brown with broad band of gold spots across
forewing and checkered fringe

**Below:** brown; forewing as above but muted; hindwing
mottled dark brown at base with distinct broad white
marginal patch

**Sexes:** similar

**Egg:** cream, laid singly on host leaves

**Larva:** dark green with a dark dorsal stripe, numerous
tiny pale yellow dots and a thin, brownish orange lateral stripe; black head

**Larval Host Plants:** primarily beggarweeds, but also
occasionally uses bush clover and Horseflyweed

**Habitat:** open, sandy woodlands, forest edges and adjacent disturbed, brushy areas

**Broods:** two generations

**Abundance:** occasional

**Compare:** Silver-Spotted Skipper (pg. 233) is larger, has
a clear white median ventral hindwing patch and
round, stubby tail.

Resident

Jan. Feb. Mar. Apr. May June July Aug. Sept. Oct. Nov. Dec.

male

Dorsal (above)
gold band

Ventral (below)
pale outer portion

outer half hoary
white

**205**

Larva

**Comments:** Delicately beautiful, the Georgia Satyr tends to be less frequently encountered than most of its common relatives. It occurs in spotty, highly localized colonies and may be easily overlooked Adults have a low, weak flight and bob slowly among the tall grasses and surrounding vegetation. Little detailed information is available about the biology and behavior of the species including the exact larval hosts utilized in the wild.

# Georgia Satyr
*Neonympha areolata*

**Family/Subfamily:** Brush-foots (Nymphalidae)/
Satyrs and Wood Nymphs (Satyrinae)

**Wingspan:** 1.40–1.75" (3.6–4.4 cm)

**Above:** uniform, unmarked brown

**Below:** brown with wavy reddish orange lines through
both wings; hindwing has a row of very elongated yellow-rimmed eyespots encircled in a thin reddish
orange oval

**Sexes:** similar

**Egg:** pale green to yellow, laid singly on host leaves

**Larva:** green with narrow light stripes, two short tails
and two reddish horns on the head

**Larval Host Plants:** currently undocumented; various
sedges likely, will accept grasses in captivity

**Habitat:** open, moist, grassy areas, wet meadows, pine
savannas, roadsides

**Broods:** two generations

**Abundance:** occasional

**Compare:** Carolina Satyr (pg. 157), Little Wood Satyr
(pg. 215) and Gemmed Satyr (pg. 187) lack the distinct
red orange ventral markings and elongated hindwing
eyespots. Mitchell's Satyr (pg. 203) has more rounded
ventral hindwing eyespots.

Resident

Jan. Feb. Mar. Apr. May June July Aug. Sept. Oct. Nov. Dec.

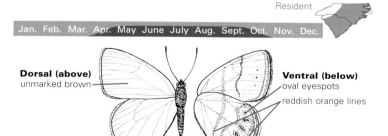

**Dorsal (above)**
unmarked brown

**Ventral (below)**
oval eyespots
reddish orange lines

Ventral

Larva

**Comments:** Although only recently established in south
Florida, the Dorantes Skipper is now a common butterfly of forest edges and adjacent open, disturbed sites
there. It is an uncommon vagrant to South Carolina.
Adults have a quick, darting flight and are frequently
drawn to available flowers. It is often abundant in
home gardens. Like several other members of the
family, the larvae construct individual shelters on the
host by weaving together one or more leaves with silk.

# Dorantes Skipper
*Urbanus dorantes*

**Family/Subfamily:** Skippers (Hesperiidae)/
Spread-wing Skippers (Pyrginae)

**Wingspan:** 1.4–1.8" (3.6–4.6 cm)

**Above:** brown with several semitransparent spots on
forewing and long hindwing tail

**Below:** brown; hindwing has two diffuse dark brown
bands, giving an overall mottled pattern

**Sexes:** similar

**Egg:** green, laid singly on the leaves of host

**Larva:** yellow-green with pinkish hue and a black head;
body has a thin dark dorsal stripe and lateral band of
dark-rimmed pale yellow spots

**Larval Host Plants:** various legumes including beggarweeds

**Habitat:** open woodlands, roadsides, old fields, fallow
agricultural land, utility easements, forest edges and
gardens

**Broods:** multiple generations

**Abundance:** rare

**Compare:** Long-tailed Skipper (pg. 223) has a dorsal
green iridescence on body and wing bases.

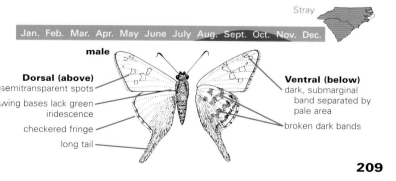

Stray

| Jan. | Feb. | Mar. | Apr. | May | June | July | Aug. | Sept. | Oct. | Nov. | Dec. |

**male**

**Dorsal (above)**
semitransparent spots
wing bases lack green
iridescence
checkered fringe
long tail

**Ventral (below)**
dark, submarginal
band separated by
pale area
broken dark bands

**209**

Male

Larva

**Comments:** Leonard's Skipper is a large, dark brown skipper with a distinctive white spot band on the hind-wing below. A late season species, it produces a single generation each year that emerges in mid to late August and continues flying well into September. Generally uncommon in the Carolinas, it may be encountered in large open fields that support tall grasses. Adults have a powerful and fast flight and tend to be quite wary and difficult to closely approach even when nectaring at many late summer blossoms.

# Leonard's Skipper
*Hesperia leonardus*

**Family/Subfamily:** Skippers (Hesperiidae)/
Banded Skippers (Hesperiinae)

**Wingspan:** 1.50–1.75" (3.8–4.4 cm)

**Above:** tawny orange with broad dark brown borders
and several small orange spots; forewings elongated
and pointed

**Below:** hindwing is reddish brown with a distinct row of
cream white spots

**Sexes:** similar, although female is primarily brown with
pale forewing spots

**Egg:** whitish green, laid singly on or near host leaves

**Larva:** olive green with a black head marked with cream

**Larval Host Plants:** a variety of grasses including bentgrass, panic grasses, and Poverty Oatgrass

**Habitat:** open, grassy areas including old fields, roadsides, wet meadows, woodland clearings and margins

**Broods:** single generation

**Abundance:** rare to uncommon

**Compare:** Indian Skipper (pg. 297) has a yellow orange
ventral hindwing with a pale spot band.

Resident

Jan. Feb. Mar. Apr. May June July Aug. Sept. Oct. Nov. Dec.

**Dorsal (above)**
tawny orange
scaling

broad dark brown
borders

**Ventral (below)**
reddish brown
hindwing

distinct row of cream
white spots

**211**

**Comments:** Freshly emerged Ocola Skippers often have a noticeable purplish sheen to the hindwings below. The butterfly can be encountered in a wide range of habitats including suburban gardens. Adults have a fast, darting flight typically within a few feet of the ground. A year-round resident of the southeast, it regularly expands its range northward each year and is particularly abundant in the fall.

# Ocola Skipper
*Panoquina ocola*

**Family/Subfamily:** Skippers (Hesperiidae)/
Banded Skippers (Hesperiinae)

**Wingspan:** 1.50–1.75" (3.8–4.4 cm)

**Above:** forewings are long and slender with pale median
spots

**Below:** brown with light veins and a subtle purple sheen;
typically darker toward margin; often has faint postmedian spot band

**Sexes:** similar

**Egg:** green, laid singly on host leaves

**Larva:** light green with yellow stripes and green head

**Larval Host Plants:** various grasses including Southern
Cutgrass

**Habitat:** marshes, pond margins, forest edges, roadsides, old fields, utility easements and gardens

**Broods:** multiple generations

**Abundance:** occasional to common

**Compare:** Salt Marsh Skipper (pg. 145) is lighter brown
beneath with distinct cream dash in center of ventral
hindwing. Brazilian Skipper (pg. 229) is larger and has a
row of semitransparent spots on ventral hindwing.

Resident Visitor

Jan. Feb. Mar. Apr. May June July Aug. Sept. Oct. Nov. Dec.

**Dorsal (above)**
long, narrow forewing
pale spots

**Ventral (below)**
darker brown toward
margin; often with
faint spot band;
subtle purple sheen
on freshly emerged
adults

213

Ventral

Larva

**Comments:** The Little Wood Satyr is one of our most abundant and commonly encountered satyrs. It dances along the forest floor with a relatively slow, bobbing flight, but can move rapidly if disturbed. Adults periodically perch on leaf litter or low vegetation with their wings partially open. Adults feed at sap flows, animal dung, rotting fungi and fermenting fruit and do not visit flowers. Unlike the similar Viola's Wood Satyr, it has several successive generations each year.

# Little Wood Satyr
*Megisto cymela*

**Family/Subfamily:** Brush-foots (Nymphalidae)/ Satyrs and Wood Nymphs (Satyrinae)

**Wingspan:** 1.5–1.9" (3.3–4.8 cm)

**Above:** brown; forewing has two prominent yellow-rimmed eyespots; hindwing has one to three (usually one is quite small) prominent yellow-rimmed eyespots

**Below:** light brown with two dark brown lines across both wings; each wing has some pearly silver markings between two large, yellow-rimmed eyespots

**Sexes:** similar, although female has larger eyespots

**Egg:** green, laid singly on host leaves

**Larva:** brown with a dark dorsal stripe, brown lateral dashes, two short stubby tails on the rear and two small horns on the head

**Larval Host Plants:** various grasses including St. Augustine Grass, bluegrass and Orchardgrass

**Habitat:** woodlands, forest clearings and margins

**Broods:** two to three generations

**Abundance:** common

**Compare:** Viola's Wood Satyr (pg. 227) is larger, has larger yellow-rimmed black eyespots, and is restricted to the southern portion of South Carolina.

Resident

Jan. Feb. Mar. Apr. May June July Aug. Sept. Oct. Nov. Dec.

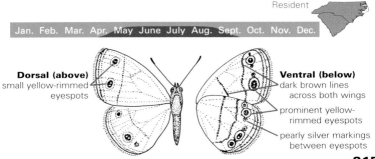

**Dorsal (above)**
small yellow-rimmed eyespots

**Ventral (below)**
dark brown lines across both wings

prominent yellow-rimmed eyespots

pearly silver markings between eyespots

**215**

Male

Female

Larva

**Comments:** A common spring species, Juvenal's Duskywing darts up and down sunlit trails and explores adjacent open sites with a quick, low flight. It frequently visits flowers or basks on bare ground with wings spread. Males perch on low vegetation and actively pursue passing objects. They occasionally puddle at damp sand or gravel. The larvae construct individual shelters on the host by folding over leaves with silk.

# Juvenal's Duskywing
*Erynnis juvenalis*

**Family/Subfamily:** Skippers (Hesperiidae)/ Spread-wing Skippers (Pyrginae)

**Wingspan:** 1.5–1.9" (3.8–4.8 cm)

**Above:** dark brown; forewing has small cluster of tiny clear spots near wingtip and one at end of cell, and is heavily patterned with brown, gray, black and tan; female has increased gray scaling and heavier pattern

**Below:** brown, lightening toward wing margin; hindwing has two small, light spots along leading margin

**Sexes:** similar, although female is lighter and more heavily patterned with larger forewing spots

**Egg:** pale green, laid singly on host leaves

**Larva:** pale green with thin, light lateral stripe and reddish brown head; head capsule has a row of light orange spots around the margin

**Larval Host Plants:** various oaks including Live Oak, Turkey Oak, Water Oak, Myrtle Oak and White Oak

**Habitat:** oak woodlands, forest edges, woodland clearings, roadsides and utility easements

**Broods:** single generation

**Abundance:** occasional to abundant

**Compare:** Horace's Duskywing (pg. 189) lacks two light spots on ventral hindwing and extensive gray scaling on dorsal forewing.

Resident

Jan. Feb. Mar. Apr. May June July Aug. Sept. Oct. Nov. Dec.

**male**

**Dorsal (above)**
small glassy spots

glassy spot at end of cell

heavy gray scaling

**Ventral (below)**
two round light spots along leading margin

**217**

Male

Ventral

Female

**Comments:** This is a common dark duskywing of the Deep South becoming increasingly rare northward. Extremely similar in appearance to the Wild Indigo Duskywing, close and careful observation is needed to ensure proper field identification but the exact determination of many (particularly worn) individuals may not be possible. Like most duskywings, the adults have a rapid, erratic flight usually within a few feet of the ground. They hold their wings open when perching or feeding. Larvae from last generation overwinter in individual leaf shelters.

# Zarucco Duskywing
*Erynnis zarucco*

**Family/Subfamily:** Skippers (Hesperiidae)/ Spread-wing Skippers (Pyrginae)

**Wingspan:** 1.6–1.9" (4.1–4.8 cm)

**Above:** dark brown; forewing has several small glassy spots toward the apex and a distinctive pale brown patch at the end of the cell; small cell spot typically lacking

**Below:** dark brown with numerous light and dark spots

**Sexes:** similar, although female is lighter with more heavily patterned wings and larger glassy spots

**Egg:** yellow green, laid singly on host leaves

**Larva:** pale green with numerous tiny white tubercles, dark dorsal stripe, yellowish white lateral stripe; dark brown head with orange spots around the margin

**Larval Host Plants:** a variety of plants in the bean family including Carolina Indigo, Hairy Indigo, Black Locust, American Wisteria, vetch and milk peas

**Habitat:** open upland woodland, forest edges and clearings, roadsides, utility easements and old fields

**Broods:** two to three generations

**Abundance:** occasional to common

**Compare:** Wild Indigo Duskywing (pg. 191) has a faint (usually) cell-end bar on the hindwing above.

Resident

Jan. Feb. Mar. Apr. May June July Aug. Sept. Oct. Nov. Dec.

**Dorsal (above)**
very dark forewing

pale patch at end of cell

**Ventral (below)**

Larva

**Comments:** Aptly named, this robust skipper had a solid golden yellow band across each forewing. Generally rare and extremely localized throughout most of its range, the Golden-Banded Skipper is most often encountered singly or in very low numbers. Adults have a somewhat slow but highly erratic, darting flight close to the ground and can be a challenge to follow. Males perch on low vegetation with their wings out-stretched.

# Golden-banded Skipper
*Autochton cellus*

**Family/Subfamily:** Skippers (Hesperiidae)/
Spread-wing Skippers (Pyrginae)

**Wingspan:** 1.5–2.0" (3.8–5.1 cm)

**Above:** brown; forewing has a golden yellow band
across the center and a white spot below the apex;
checkered fringe

**Below:** forewing is marked as above, but paler and mottled with faint dark brown spots; hindwing has two
irregular dark brown bands and a light gray frosted
patch along the margin

**Sexes:** similar

**Egg:** yellow, laid in short strings on host leaves

**Larva:** yellow green with a broad yellow lateral stripe,
numerous small yellow dots; reddish brown head has
two round yellow spots on the lower half

**Larval Host Plants:** primarily beggarweeds, occasionally uses bush clover and Horseflyweed

**Habitat:** moist woodlands, stream corridors, wetland
margins

**Broods:** two generations

**Abundance:** occasional

**Compare:** Hoary Edge (pg. 205) has a broad, dirty white
patch along the margin of the ventral hindwing.

Resident

Jan. Feb. Mar. Apr. May June July Aug. Sept. Oct. Nov. Dec.

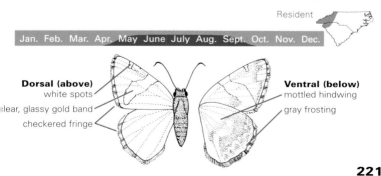

**Dorsal (above)**
white spots
clear, glassy gold band
checkered fringe

**Ventral (below)**
mottled hindwing
gray frosting

Ventral

Larva

**Comments:** Resembling a small swallowtail, the Long-tailed Skipper is one of the most common and distinctive skippers in the Carolinas. Adults have a quick, low flight. They are fond of flowers and often abundant in home gardens. The larvae construct individual shelters on the host by folding over small sections of a leaf with silk. Older larvae may use the entire leaf or connect several leaves together. The butterfly is migratory, and moves southward each fall to overwinter in warmer portions of Florida.

# Long-tailed Skipper
*Urbanus proteus*

**Family/Subfamily:** Skippers (Hesperiidae)/
Spread-wing Skippers (Pyrginae)

**Wingspan:** 1.5–2.0" (3.8–5.1 cm)

**Above:** brown with iridescent blue-green scaling on wing
bases and body; long hindwing tail; forewing has several semitransparent spots

**Below:** brown; hindwing has two crisp dark brown
bands; forewing has a continuous brown submarginal
band

**Sexes:** similar

**Egg:** pale yellow, laid singly on the underside of host
leaves

**Larva:** yellow-green with a dark dorsal stripe, yellow lateral stripe and crimson head; body is covered with tiny
black spots

**Larval Host Plants:** a wide variety of legumes including beggarweeds, wisteria, beans and Kudzu

**Habitat:** open, disturbed sites including roadsides, old
fields, fallow agricultural land, utility easements, forest
edges and gardens

**Broods:** multiple generations

**Abundance:** common to abundant

**Compare:** Dorantes Skipper (pg. 209) lacks iridescent
blue-green dorsal coloration.

Visitor

Jan. Feb. Mar. Apr. May June July Aug. Sept. Oct. Nov. Dec.

**Dorsal (above)**
semitransparent
spots

iridescent
blue-green

blue-green body

**Ventral (below)**
continuous brown
submarginal band

distinct dark bands
and spots

long tail

**223**

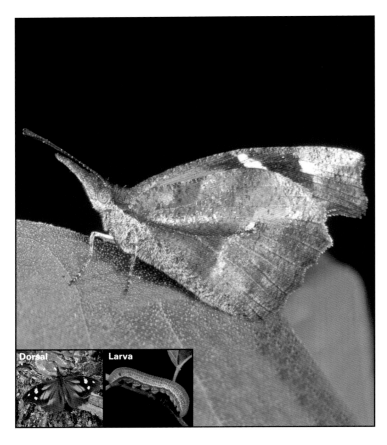

Dorsal

Larva

**Comments:** The American Snout gets its name from the unusually long labial palpi that resemble an elongated nose. This unique feature, combined with the cryptic coloration of the wings beneath enhances the butterfly's overall "dead leaf" appearance when at rest. Adults have a rapid, somewhat erratic flight and are commonly drawn to flowers. Males readily puddle at damp ground. This butterfly typically rests and feeds with its wings closed, so its dorsal color is seen primarily in flight or when the butterfly is basking in the sun.

# American Snout
*Libytheana carinenta*

**Family/Subfamily:** Brush-foots (Nymphalidae)/ Snouts (Libytheinae)

**Wingspan:** 1.6–1.9" (4.1–4.8 cm)

**Above:** brown with orange patches and white forewing spots; forewing apex is extended and squared off

**Below:** brown with orange basal forewing scaling and white spots; hindwing variable; plain gray brown or pinkish brown with heavy mottling

**Sexes:** similar

**Egg:** tiny white eggs laid in axils of host leaves

**Larva:** light green with numerous small yellow dots and yellow lateral stripe; rear portion has two small black lateral spots

**Larval Host Plants:** Common Hackberry and Sugarberry

**Habitat:** rich, deciduous woodlands, stream corridors, forest edges, woodland clearings and adjacent open brushy areas

**Broods:** multiple generations

**Abundance:** occasional

**Compare:** unique

Resident

Jan. Feb. Mar. Apr. May June July Aug. Sept. Oct. Nov. Dec.

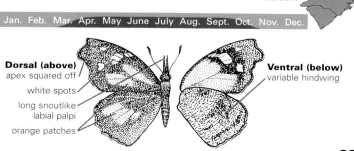

**Dorsal (above)**
apex squared off
white spots
long snoutlike labial palpi
orange patches

**Ventral (below)**
variable hindwing

**225**

**Comments:** There remains considerable debate about the taxonomic status of this attractive woodland butterfly. Some treat it as a separate species, while others feel it is best considered a form of the widespread Little Wood Satyr. It scurries along the forest floor with a quick, bobbing flight and periodically lands on leaf litter or low vegetation. Like most satyrs, it does not visit flowers but instead is drawn to sap flows, dung, rotting fungi and fermenting fruit. The large yellow-rimmed eyespots presumably help deflect attack away from its vulnerable body.

# Viola's Wood Satyr
*Megisto viola*

**Family/Subfamily:** Brush-foots (Nymphalidae)/
Satyrs and Wood Nymphs (Satyrinae)

**Wingspan:** 1.6–2.0" (4.1–5.1 cm)

**Above:** brown with large yellow-rimmed eyespots on
both wings; male has two conspicuous eyespots on
the forewing and one to two on the hindwing; female
has two on the hindwing

**Below:** light brown with two dark brown lines and large,
yellow-rimmed eyespots; silver markings between
eyespots

**Sexes:** similar, although female has larger eyespots

**Egg:** green, laid singly on host leaves

**Larva:** brown with dark dorsal stripe and two stubby tails

**Larval Host Plants:** various grasses including St.
Augustine Grass

**Habitat:** shady woodlands, forest edges and adjacent
open areas

**Broods:** single spring generation

**Abundance:** occasional to common

**Compare:** Little Wood Satyr (pg. 215) is smaller and has
smaller yellow-rimmed eyespots.

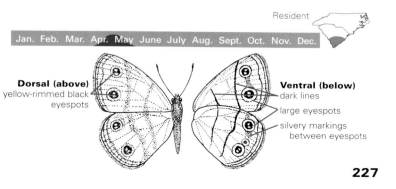

Resident

Jan. Feb. Mar. Apr. May June July Aug. Sept. Oct. Nov. Dec.

**Dorsal (above)**
yellow-rimmed black
eyespots

**Ventral (below)**
dark lines

large eyespots

silvery markings
between eyespots

Dorsal

Larva

**Comments:** The Brazilian Skipper is also referred to as the Canna Skipper and can be a common nuisance or aesthetic pest on ornamental cannas. The transparent, alien-looking larvae construct individual shelters by rolling over sections of the host leaf with silk. Inside they hide safely out of the sight of would-be predators and come out to feed mainly at night. The adults have a strong, rapid flight, usually low to the ground, and visit a wide range of flowers. Primarily a tropical butterfly, it is a seasonal Carolinas colonist and unable to survive freezing temperatures in any life stage.

## Brazilian Skipper
*Calpodes ethlius*

**Family/Subfamily:** Skippers (Hesperiidae)/
Banded Skippers (Hesperiinae)

**Wingspan:** 1.75–2.25" (4.4–5.7 cm)

**Above:** brown; elongated forewing is lighter at base with
several semitransparent spots across the center; hindwing has postmedian band of three small
semitransparent spots and tapers toward bottom

**Below:** as above but lighter brown

**Sexes:** similar

**Egg:** gray-green, laid singly on host leaves

**Larva:** green, semitransparent with orange-brown head;
head has a central black dot

**Larval Host Plants:** Alligator Flag, Indian Shot, ornamental cannas

**Habitat:** marshes, pond edges, parks, gardens and various urban areas

**Broods:** multiple generations

**Abundance:** occasional to common

**Compare:** Ocola Skipper (pg. 213) is smaller and lacks
ventral hindwing band of semitransparent spots.

Visitor

Jan. Feb. Mar. Apr. May June July Aug. Sept. Oct. Nov. Dec.

**male**

**Dorsal (above)**
long, narrow forewing
large translucent
spots
tapers to stubby lobe

**Ventral (below)**
band of
semitransparent
spots

**229**

Ventral    Larva

**Comments:** The Appalachian Brown is a medium-sized brown butterfly. Often spotty and localized, it prefers moist, shaded habitats. Adults have an erratic, bouncing flight and travel low through wetland vegetation, stopping frequently to perch. Even when disturbed, they typically fly only a short distance before alighting again. As a result of their reclusive behavior, active colonies may be easily overlooked. Like most satyrs, the adults feed at sap flows, fermenting fruit, animal dung or rotting fungi and do not visit flowers.

# Appalachian Brown
*Satyrodes appalachia*

**Family/Subfamily:** Brush-foots (Nymphalidae)/
Satyrs and Wood Nymphs (Satyrinae)

**Wingspan:** 1.90–2.25" (4.8–5.7 cm)

**Above:** light brown with small, solid black eyespots

**Below:** soft brown; forewing has a row of four double-rimmed black eyespots, the middle two generally smaller than one above or below; hindwing has a row of five to six double-rimmed black eyespots with pale centers bordered inwardly by a dark sinuous post-median line

**Sexes:** similar, although female is generally paler brown with larger eyespots

**Egg:** greenish white, laid singly on or near host leaves

**Larva:** light green with narrow longitudinal yellow stripes, two short tails on the rear and two reddish horns on the head

**Larval Host Plants:** various grasses and sedges including Upright Sedge and Fowl Mannagrass

**Habitat:** wooded swamps, moist, grassy glades, stream corridors and forest margins

**Broods:** two generations

**Abundance:** occasional; localized

**Compare:** unique

Resident

Jan. Feb. Mar. Apr. May June July Aug. Sept. Oct. Nov. Dec.

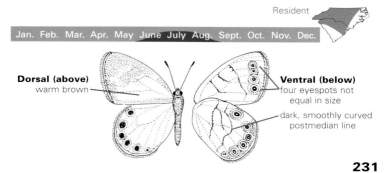

**Dorsal (above)**
warm brown

**Ventral (below)**
four eyespots not
equal in size

dark, smoothly curved
postmedian line

**231**

Larva

**Comments:** The Silver-spotted Skipper is a large, robust butterfly with a rapid, darting flight. It is a common garden visitor. They have a long proboscis and can easily gain access to nectar from a variety of flowers. Males perch on shrubs or overhanging branches and aggressively investigate passers-by. The colorful larvae construct individual shelters on the host by tying one or more leaves together with silk.

# Silver-spotted Skipper
*Epargyreus clarus*

**Family/Subfamily:** Skippers (Hesperiidae)/ Spread-wing Skippers (Pyrginae)

**Wingspan:** 1.75–2.40" (4.4–6.1 cm)

**Above:** brown with median row of gold spots on forewing and checkered wing fringe; hindwing is tapered into small, rounded, lobe-like tail

**Below:** brown; forewing as above; hindwing has distinct, elongated clear silver-white patch in center

**Sexes:** similar

**Egg:** green, laid singly on host leaves

**Larva:** yellow-green with dark bands and reddish brown head

**Larval Host Plants:** wide variety of legumes including Black Locust, wisteria, bush clover, False Indigo, Kudzu, Groundnut, Butterfly Pea and beggarweeds

**Habitat:** forest edges, open woodlands, roadsides, utility easements, old fields and gardens

**Broods:** multiple generations

**Abundance:** occasional to common

**Compare:** Hoary Edge (pg. 205) is smaller and has marginal white ventral hindwing patch.

Resident

Jan. Feb. Mar. Apr. May June July Aug. Sept. Oct. Nov. Dec.

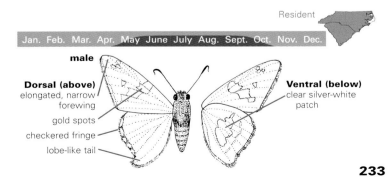

male

**Dorsal (above)**
elongated, narrow forewing
gold spots
checkered fringe
lobe-like tail

**Ventral (below)**
clear silver-white patch

**233**

Female

Ventral  Larva

**Comments:** The Common Buckeye is one of our most distinctive butterflies. The large eyespots help deflect attack away from its vulnerable body and may also startle predators. It is fond of open, sunny locations with low-growing vegetation and may be an occasional garden visitor. Adults frequently alight on bare soil or gravel but are extremely wary and difficult to approach. Flight is rapid and low to the ground. Unable to withstand freezing temperatures, the Buckeye annually undertakes a southward fall migration and overwinters in warmer Gulf Coast locations.

# Common Buckeye
*Junonia coenia*

**Family/Subfamily:** Brush-foots (Nymphalidae)/
True Brush-foots (Nymphalinae)

**Wingspan:** 1.5–2.7" (3.8–6.9 cm)

**Above:** brown with prominent eyespots; forewing bears a distinct white patch and two small orange bars

**Below:** forewing has prominent white band; hindwings seasonally variable in color; summer forms are light brown with numerous pattern elements; cool-season forms are reddish brown with reduced markings

**Sexes:** similar, although female has broader wings and larger hindwing eyespots

**Egg:** dark green, laid singly on host leaves

**Larva:** black with lateral white stripes, orange patches and branched spines

**Larval Host Plants:** plants in many families including toadflax, False Foxglove, frogfruit, plantain, twinflower and wild petunia

**Habitat:** fields, pastures, roadsides, fallow agricultural land, gardens, open pineland, disturbed sites

**Broods:** multiple generations, overwinters as adult

**Abundance:** occasional to locally common

**Compare:** unique

Resident

Jan. Feb. Mar. Apr. May June July Aug. Sept. Oct. Nov. Dec.

**Dorsal (above)**
white band that
surrounds
eyespot

orange bars

eyespots

orange band

**Ventral (below)**
seasonally variable

**235**

Male

Female

Ventral

**Comments:** Although the Cofaqui Giant-Skipper is somewhat dwarfed by its larger relative, the Yucca Giant-Skipper, it is still an impressive butterfly. It is an inhabitant of open pine woodlands that support its larval host. Larvae burrow into the host stem and feed on the root. Larvae overwinter. Adults have a rapid, powerful flight that produces a noticeable buzz when close. Males perch on vegetation along trails or clearings and actively defend their territory. It is rare and seldom-seen throughout its limited range. Witnessing even one individual will make your day!

# Cofaqui Giant-Skipper
*Megathymus cofaqui*

**Family/Subfamily:** Skippers (Hesperiidae)/
Giant-Skippers (Megathyminae)

**Wingspan:** 1.9–2.4" (4.8–6.1 cm)

**Above:** dark brown; male has a broad yellow forewing
band and pale hindwing border

**Below:** dull gray brown with variable small white spots
and light gray scaling toward outer margin

**Sexes:** similar, although female is larger with broader
wings and a postmedian row of small yellow spots on
the hindwing

**Egg:** cream, laid singly on host leaves

**Larva:** cream with a brownish black head

**Larval Host Plants:** yuccas including Spanish Dagger
and Stiff-leaved Bear Grass

**Habitat:** dry woodlands, pinelands and scrub

**Broods:** two generations

**Abundance:** rare to occasional

**Compare:** Yucca Giant-Skipper (pg. 249) is larger with
more pointed and elongated forewings, a narrower
straight yellow forewing band above and darker ventral
wings with extensive frosting.

Resident

Jan. Feb. Mar. Apr. May June July Aug. Sept. Oct. Nov. Dec.

**male**

**Dorsal (above)**
broad yellow band
arches inward

pale border

**Ventral (below)**
gray brown with
scattered white
spots

**237**

Larva

**Comments:** Although called the Northern Pearly Eye, the range of this butterfly extends into the Deep South, brushing the western edge of the Carolinas. It occurs in localized, spotty colonies in association with its larval host, but may be common when encountered. Adults have a quick, bobbing flight and maneuver close to the ground. They frequently alight on low vegetation, tree trunks or on leaf litter. They feed at sap flows, rotting fruit, decaying vegetation, fungi and dung. Unlike most butterflies, the adults are active even on overcast days and often fly late into the evening.

# Northern Pearly Eye
*Enodia anthedon*

**Family/Subfamily:** Brush-foots (Nymphalidae)/
Satyrs and Wood Nymphs (Satyrinae)

**Wingspan:** 1.75–2.60" (4.3–6.6 cm)

**Above:** light brown with black eyespots in a pale field;
hindwing has a slightly scalloped margin; antennal
clubs are black at base

**Below:** brown with a violet cast; forewing has a straight
row of four yellow-rimmed dark eyespots; hindwing
has a cream band enclosing a row of yellow-rimmed
dark eyespots with light highlights

**Sexes:** similar, although female generally has broader,
more rounded wings and larger eyespots

**Egg:** greenish white, laid singly on host leaves

**Larva:** yellow green with narrow longitudinal yellow
stripes, a dark green dorsal stripe, two short red-tipped
tails on the rear and two reddish horns on the head

**Larval Host Plants:** various grasses including
Whitegrass, Indian Woodoats, Silver Plumegrass,
Bearded Shorthusk and Reed Canarygrass

**Habitat:** moist, shady woods, stream corridors, marsh
edges and semi-open grassy areas along forest edges

**Broods:** two generations

**Abundance:** occasional; localized

**Compare:** Southern Pearly Eye (pg. 245) has orange-
tipped antennae.

Resident

Jan. Feb. Mar. Apr. May June July Aug. Sept. Oct. Nov. Dec.

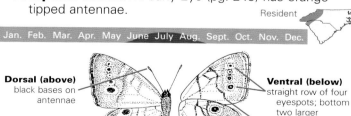

**Dorsal (above)**
black bases on
antennae

**Ventral (below)**
straight row of four
eyespots; bottom
two larger

dark, gently curved
postmedian line

Male

Ventral

Larva

**Comments:** The Hackberry Butterfly is named for its common larval host. Though it prefers shady woodlands, it may show up in suburban yards or parks. It has a strong, rapid flight and perches on sunlit leaves, overhanging branches or tree trunks along forest trails and woodland edges. Adults are exceedingly inquisitive and dart out to investigate passing objects, sometimes even landing on humans. They are drawn to sap flows or rotting fruit and do not visit flowers. Although often spotty and localized, the species can be quite abundant when encountered.

# Hackberry Butterfly
*Asterocampa celtis*

**Family/Subfamily:** Brush-foots (Nymphalidae)/ Emperors (Apaturinae)

**Wingspan:** 2.0–2.6" (5.1–6.6 cm)

**Above:** amber-brown with dark markings and borders; forewing bears several small white spots near the apex and a single submarginal black eyespot; hindwing has a postmedian row of dark spots

**Below:** as above with muted coloration; hindwing has postmedian row of yellow-rimmed black spots with blue green centers

**Sexes:** similar, although female has broader wings

**Egg:** cream-white, laid singly or in small clusters on leaves

**Larva:** light green with two narrow dorsal yellow stripe; mottled with small yellow spots; dark head bears two stubby, branched horns; rear end has a pair of short tails

**Larval Host Plants:** Common Hackberry, Sugarberry and Dwarf Hackberry

**Habitat:** moist, rich woodlands, forest margins and clearings, stream corridors, parks and yards

**Broods:** multiple generations

**Abundance:** occasional to common

**Compare:** Tawny Emperor (pg. 247) is more orange-brown above and lacks white forewing spots and single forewing eyespot.

Resident

Jan. Feb. Mar. Apr. May June July Aug. Sept. Oct. Nov. Dec.

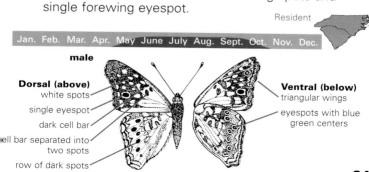

**male**

**Dorsal (above)**
white spots
single eyespot
dark cell bar
ell bar separated into two spots
row of dark spots

**Ventral (below)**
triangular wings
eyespots with blue green centers

**241**

**Comments:** The Common Wood Nymph is a medium-sized butterfly primarily encountered in open, grassy meadows and fields, though it may also be found in forest clearings and margins. Adults have a low, relaxed flight and bob erratically through the vegetation. Well camouflaged when resting, it can be a challenge to locate or follow. Unlike most satyrs, it is an opportunistic feeder and frequently visits flowers along with sap flows and fermenting fruit.

# Common Wood Nymph
*Cercyonis pegala*

**Family/Subfamily:** Brush-foots (Nymphalidae)/ Satyrs and Wood Nymphs (Satyrinae)

**Wingspan:** 1.8–2.8" (4.6–7.1 cm)

**Above:** brown; forewing has large postmedian yellow patch containing two dark eyespots

**Below:** brown with dark striations; forewing has large postmedian yellow patch containing two dark eyespots; hindwing has postmedian row of small, yellow-rimmed dark eyespots

**Sexes:** similar, although female is paler and has larger eyespots

**Egg:** cream, laid singly on host leaves

**Larva:** green with dark green dorsal stripe and light side stripes

**Larval Host Plants:** various grasses including bluegrass, Poverty Oatgrass and Purpletop Grass

**Habitat:** open woodlands, forest edges, marshes, grassy fields and roadsides

**Broods:** single generation

**Abundance:** occasional to common

**Compare:** unique

Resident

Jan. Feb. Mar. Apr. May June July Aug. Sept. Oct. Nov. Dec.

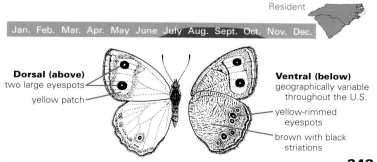

**Dorsal (above)**
two large eyespots
yellow patch

**Ventral (below)**
geographically variable throughout the U.S.
yellow-rimmed eyespots
brown with black striations

**243**

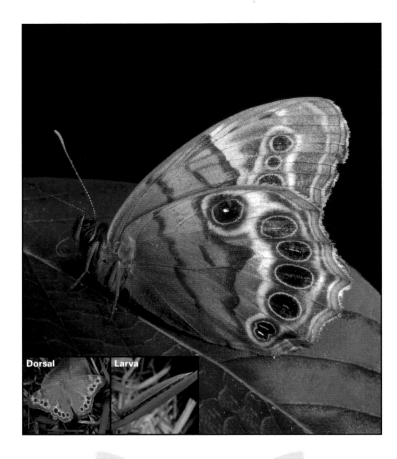

Dorsal | Larva

**Comments:** The Southern Pearly Eye is one of our largest satyrs. It prefers moist shady areas and is typically found in close association with its larval host. The butterfly has a spotty, localized distribution but may be common when found. Adults have a quick, bobbing flight and frequently alight on low vegetation or tree trunks. They do not visit flowers, but instead prefer to feed at sap flows, rotting fruit, decaying vegetation and dung. Unlike most butterflies, adults are often active on overcast days and fly until dusk.

# Southern Pearly Eye
*Enodia portlandia*

**Family/Subfamily:** Brush-foots (Nymphalidae)/
Satyrs and Wood Nymphs (Satyrinae)

**Wingspan:** 2.00–2.75" (5.1–7.0 cm)

**Above:** light brown; forewing has dark eyespots in light
postmedian patch

**Below:** purplish brown; hindwing has submarginal cream
band and row of yellow-rimmed dark eyespots

**Sexes:** similar, although female is generally somewhat
lighter in color

**Egg:** pale green, laid singly on host leaves

**Larva:** green with narrow light stripes, two short tails
and two reddish horns on the head

**Larval Host Plants:** Giant Cane and Switchcane

**Habitat:** moist, shaded woodlands, stream corridors and
swamp margins

**Broods:** multiple generations

**Abundance:** occasional to locally common

**Compare:** Creole Pearly Eye (pg. 251) has five yellow-
rimmed black eyespots and a dark line through the
center of the ventral forewing that bulges outward
near costal margin.

Resident

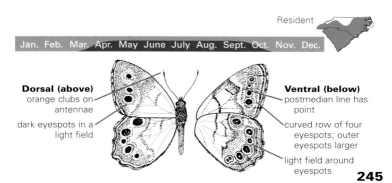

Jan. Feb. Mar. Apr. May June July Aug. Sept. Oct. Nov. Dec.

**Dorsal (above)**
orange clubs on
antennae

dark eyespots in a
light field

**Ventral (below)**
postmedian line has
point

curved row of four
eyespots; outer
eyespots larger

light field around
eyespots

**245**

Male

Male | Female | Larva

**Comments:** The Tawny Emperor is superficially similar to the Hackberry Butterfly with which it flies. It is seldom found far from stands of its larval hosts. Adults are rapid, strong fliers and often difficult to approach. Males perch on sunlit leaves or on the sides of large trees and dart out quickly to investigate passing objects, occasionally alighting on humans. The developing larvae remain together and feed communally through the first three instars before becoming more solitary.

# Tawny Emperor
*Asterocampa clyton*

**Family/Subfamily:** Brush-foots (Nymphalidae)/ Emperors (Apaturinae)

**Wingspan:** 2.00–2.75" (5.1–7.0 cm)

**Above:** orange-brown with dark markings and borders; hindwing has a postmedian row of dark spots

**Below:** as above with muted gray-brown cast and small, dark hindwing eyespots

**Sexes:** similar, although female is much larger with broader, rounder wings

**Egg:** cream-white, laid in large pyramidal clusters on the underside of host leaves

**Larva:** light green with broad dorsal yellow stripes, narrow yellow lateral stripes and mottled with small yellow spots; head is green and bears two stubby, branched horns; rear end has a pair of short tails

**Larval Host Plants:** Common Hackberry, Sugarberry and Dwarf Hackberry

**Habitat:** rich, moist deciduous woodlands, forest clearings and margins, stream corridors, parks and yards

**Broods:** multiple generations

**Abundance:** occasional; locally abundant

**Compare:** Hackberry Butterfly (pg. 241) is lighter brown with single black eyespot and white spots on forewing.

Resident

Jan. Feb. Mar. Apr. May June July Aug. Sept. Oct. Nov. Dec.

**male**

**Dorsal (above)**
narrow, triangular wings

no white spots

no black eyespot

two solid bars

**Ventral (below)**
muted pattern

small, dark eyespots

**247**

Ventral

**Comments:** The Yucca Giant-Skipper is a beautifully
marked butterfly with elongated wings. Adults have a
rapid, powerful flight that produces a noticeable buzz.
Males perch on vegetation along trails or clearings and
actively defend their territory. The butterfly is particularly active in late afternoon or early evening. The
larvae feed on host leaves when young but soon burrow into stem and large taproot, forming a long tunnel.

# Yucca Giant-Skipper
*Megathymus yuccae*

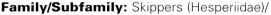

**Family/Subfamily:** Skippers (Hesperiidae)/ Giant-Skippers (Megathyminae)

**Wingspan:** 2.0–2.8" (5.1–7.1 cm)

**Above:** blackish brown with yellow hindwing border and straight band of yellow spots on forewing; small white spots near the forewing apex; female has small post-median band of tiny orange spots on hindwing

**Below:** chocolate brown with marginal frosting and distinct white patch along upper margin of hindwing

**Sexes:** similar, although female is significantly larger with broader wings and an additional row of yellow spots on hindwing

**Egg:** pinkish brown, laid singly on host leaves

**Larva:** light tan with reddish brown head

**Larval Host Plants:** yuccas including Spanish Dagger and Stiff-leaved Bear Grass

**Habitat:** dry woodlands, pinelands and scrub

**Broods:** single generation

**Abundance:** occasional

**Compare:** Cofaqui Giant-Skipper (pg. 237) has a broad, curved yellow forewing band and is duller gray-brown below with less prominent scattered small white hindwing spots.

Resident

Jan. Feb. Mar. Apr. May June July Aug. Sept. Oct. Nov. Dec.

**male**

**Dorsal (above)**
elongated forewing

straight yellow band does not reach costal margin

yellow band

**Ventral (below)**
white spot

gray frosting

Larva

**Comments:** The Creole Pearly Eye is rarer and more localized in occurrence than its two other close relatives. It seldom wanders far from stands of its larval host. It is often encountered alongside the similar Southern Pearly Eye and as a result may go undetected. Adults have a quick, bobbing flight and maneuver close to the ground through the forest understory, making them a challenge to follow. They frequently perch on tree trunks or other vegetation. They do not visit flowers but instead prefer to feed at sap flows, rotting fruit, decaying vegetation, fungi and dung.

# Creole Pearly Eye
*Enodia creola*

**Family/Subfamily:** Brush-foots (Nymphalidae)/
Satyrs and Wood Nymphs (Satyrinae)

**Wingspan:** 2.30–2.75" (5.8–7.0 cm)

**Above:** light brown; forewing has black eyespots in pale
field; male has dark, raised scent-scale patches; hindwing has a row of large, solid black eyespots and
slightly scalloped margin

**Below:** brown with a violet cast; forewing has a straight
row of five yellow-rimmed dark eyespots and a dark
postmedian line bulging outward near the costal margin

**Sexes:** similar, female lacks the scent-scale patches, and
has broader, more rounded wings and larger eyespots

**Egg:** greenish white, laid singly on host leaves

**Larva:** yellow green with narrow longitudinal yellow
stripes, two short red-tipped tails on the rear and two
reddish horns on the head

**Larval Host Plants:** Switchcane

**Habitat:** moist, shaded woodlands, dense upland
forests, stream corridors and swamp margins

**Broods:** two generations

**Abundance:** occasional; localized

**Compare:** Southern (pg. 245) and Northern Pearly Eye
(pg. 239) have four ventral forewing eyespots.

Resident

Jan. Feb. Mar. Apr. May June July Aug. Sept. Oct. Nov. Dec.

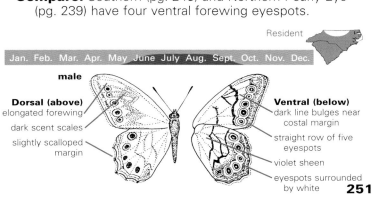

**male**

**Dorsal (above)**
elongated forewing

dark scent scales

slightly scalloped
margin

**Ventral (below)**
dark line bulges near
costal margin

straight row of five
eyespots

violet sheen

eyespots surrounded
by white

**251**

Ventral

**Comments:** During years with population outbreaks, this northland butterfly periodically emigrates to more southern locations. It should be considered a rare stray to North Carolina.

## Compton Tortoiseshell
*Nymphalis vaualbum*

**Family/Subfamily:** Brush-foots (Nymphalidae)/
True Brush-foots (Nymphalinae)

**Wingspan:** 2.5–3.1" (6.4–7.9 cm)

**Above:** rusty brown with heavy black spots and golden
scaling toward outer margin; each wing bears a bright
white spot just below the apex; forewing apex is
extended and squared off; hindwing bears a single
short, stubby tail

**Below:** appears bark-like; heavily striated with gray and
brown; outer portion noticeably lighter than basal half

**Sexes:** similar

**Egg:** green, laid in clusters on host

**Larva:** light green with pale mottling and several rows of
lateral cream and dorsal black branched spines

**Larval Host Plants:** birch, willow and Quaking Aspen

**Habitat:** deciduous or mixed forests, clearings, woodland roads, forest edges and adjacent open areas

**Broods:** single generation

**Abundance:** rare

**Compare:** unique

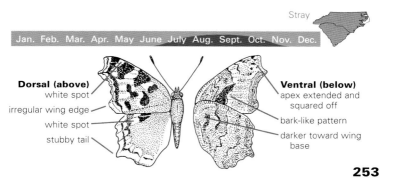

Stray

Jan. Feb. Mar. Apr. May June July Aug. Sept. Oct. Nov. Dec.

**Dorsal (above)**
white spot
irregular wing edge
white spot
stubby tail

**Ventral (below)**
apex extended and
squared off
bark-like pattern
darker toward wing
base

**253**

Male

Ventral     Larva

**Comments:** With a wingspan close to six inches, the
Giant Swallowtail is one of the largest butterflies in
North America. When nectaring, adults continuously
flutter their wings much like a hummingbird. This
behavior coupled with a long proboscis enables them
to visit a wide range of flowers, including many that
otherwise might not easily support their weight. Giant
Swallowtail larvae, often called "orange dogs"
because of their fondness for citrus, occasionally
become minor pests in commercial orange groves.

# Giant Swallowtail
*Papilio cresphontes*

**Family/Subfamily:** Swallowtails (Papilionidae)/
Swallowtails (Papilioninae)

**Wingspan:** 4.5–5.5" (11.4–14.0 cm)

**Above:** chocolate brown with broad crossing bands of
yellow spots; characteristic diagonal band extends
from tip of forewing to base of abdomen; hindwing tail
has yellow center

**Below:** cream yellow with brown markings and blue
median hindwing band

**Sexes:** similar, although female is generally larger

**Egg:** amber-brown, laid singly on upperside of host
leaves

**Larva:** brown with yellow and cream patches; resembles
bird dropping

**Larval Host Plants:** Hercules Club, Wafer Ash, Prickly
Ash and cultivated citrus

**Habitat:** woodlands, pastures, forest edges, stream corridors, open pinelands and suburban gardens

**Broods:** multiple generations

**Abundance:** occasional

**Compare:** Palamedes Swallowtail (pg. 77) has cream
markings and lacks crossing forewing bands.

Resident

Jan. Feb. Mar. Apr. May June July Aug. Sept. Oct. Nov. Dec.

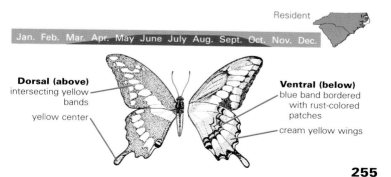

**Dorsal (above)**
intersecting yellow
bands

yellow center

**Ventral (below)**
blue band bordered
with rust-colored
patches

cream yellow wings

**255**

**Comments:** The Red-banded Hairstreak has a rapid, erratic flight. Males typically perch on sunlit leaves of shrubs or small trees (often on their hosts) and readily interact with other individuals, often spiraling high into the air before returning to a nearby perch. Unlike most other butterflies, female Red-banded Hairstreaks do not lay eggs directly on the larval host. Instead, they land on the ground below appropriate hosts and deposit the small eggs singly on underside of dead, fallen leaves or other debris. The resulting larvae feed primarily on detritus below certain hosts.

# Red-banded Hairstreak
*Calycopis cecrops*

**Family/Subfamily:** Gossamer Wings (Lycaenidae)/ Hairstreaks (Theclinae)

**Wingspan:** 0.75–1.00" (1.9–2.5 cm)

**Above:** male is slate gray above with no markings; female is slate gray with iridescent blue scaling on hindwing; hindwing bears two short tails

**Below:** light gray with broad, red band edged outwardly by a thin, wavy white line; blue scaling and a black eyespot near tails

**Sexes:** similar, although female has blue scaling above

**Egg:** cream brown, laid on dead leaves below host

**Larva:** pinkish brown with numerous short hairs

**Larval Host Plants:** Wax Myrtle, Winged Sumac, Mango, Staghorn Sumac and Fragrant Sumac; feed primarily on dead plant material below host

**Habitat:** woodland edges and adjacent disturbed, brushy areas, suburban gardens

**Broods:** multiple generations

**Abundance:** common

**Compare:** Southern Hairstreak (pg. 137) lacks complete red hindwing band.

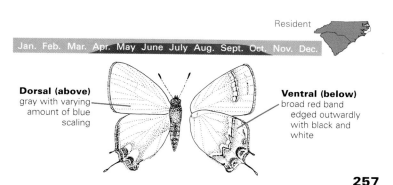

Resident

Jan. Feb. Mar. Apr. May June July Aug. Sept. Oct. Nov. Dec.

**Dorsal (above)**
gray with varying
amount of blue
scaling

**Ventral (below)**
broad red band
edged outwardly
with black and
white

Male

Larva

**Comments:** This early-season blue is easily distinguished from the superficially similar Spring Azure with which it often flies by its dark, charcoal gray-infused wings. Although now considered a distinct species, individuals were previously through to be rare dark-colored aberrations of the Spring Azure. It typically occurs in small, isolated colonies that are often best discovered by first locating patches of its larval host. Adults have a quick, directed flight usually close to the ground. Males often join other azures at mud puddles or stream banks to sip moisture. It typically rests and feeds with its wings closed; dorsal color shows in flight or while basking.

# Dusky Azure
*Celastrina nigra*

**Family/Subfamily:** Gossamer Wings (Lycaenidae)/
Blues (Polyommatinae)

**Wingspan:** 0.75–1.25" (1.9–3.2 cm)

**Above:** male is uniform dark charcoal gray; female is
pale gray blue with extensive, broad dark wing borders

**Below:** light gray with small black spots; hindwing has
pale but prominent dark zigzag band along margin
enclosing a row of small dark spots

**Sexes:** dissimilar; female is pale gray blue with extensive, broad dark wing borders, some white scaling,
and distinct narrow dark cell-end bars

**Egg:** blue gray eggs laid singly young shoots, new leaves
or flower buds of host

**Larva:** yellow green with pale lateral stripes

**Larval Host Plants:** goatsbeard or Bride's Feathers

**Habitat:** moist, cool, shaded, deciduous woodlands, forest trails, woodland roads, shaded ravines, wooded
ridgetops and stream corridors

**Broods:** single generation

**Abundance:** occasional

**Compare:** Spring Azure (pg. 87) is blue above and has
more extensive dark gray scaling below.

Resident

Jan. Feb. Mar. Apr. May June July Aug. Sept. Oct. Nov. Dec.

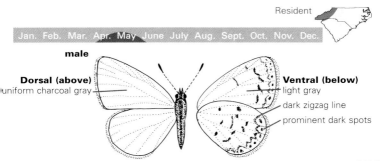

**male**

**Dorsal (above)**
uniform charcoal gray

**Ventral (below)**
light gray

dark zigzag line

prominent dark spots

**259**

Ventral

Larva

**Comments:** Widespread and abundant, the Gray
Hairstreak is one of our most commonly encountered
hairstreaks. It is extremely fond of flowers and a frequent garden visitor. The small hairlike tails on the
hindwing resemble antennae and presumably help
deflect the attack of would-be predators away from
the insect's vulnerable body. This charade, employed
by many members of the family, is enhanced by the
bright orange eyespots and converging lines on the
wings below that draw attention to this unique "false
head" feature.

# Gray Hairstreak
*Strymon melinus*

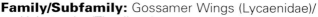

**Family/Subfamily:** Gossamer Wings (Lycaenidae)/ Hairstreaks (Theclinae)

**Wingspan:** 1.0–1.5" (2.5–3.8 cm)

**Above:** slate gray with distinct reddish orange-capped black hindwing spot above tail

**Below:** light gray with black-and-white line across both wings (often with some orange); hindwing has reddish orange-capped black spot and blue scaling above tail

**Sexes:** similar; female larger with broader wings

**Egg:** light green, laid singly on flower buds or flowers of host

**Larva:** highly variable; bright green with lateral cream stripes to pinkish red

**Larval Host Plants:** wide variety of plants including Partridge Pea, beggarweeds, milk peas, milkvetch, lupine, bush clover, clover, vetch, mallow and Sida

**Habitat:** open, disturbed sites including roadsides, fallow agricultural land, old fields and gardens

**Broods:** multiple generations

**Abundance:** common

**Compare:** White M Hairstreak (pg. 95) has red ventral hindwing spot without a black pupil, distinct zigzag white line and a small white spot along the costal margin of hindwing.

Resident

Jan. Feb. Mar. Apr. May June July Aug. Sept. Oct. Nov. Dec.

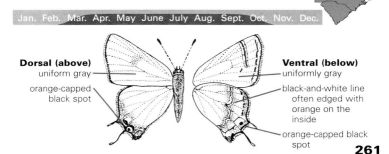

**Dorsal (above)**
uniform gray

orange-capped black spot

**Ventral (below)**
uniformly gray

black-and-white line often edged with orange on the inside

orange-capped black spot

**261**

Female

Female

**Comments:** The diminutive Early Hairstreak is one of
our scarcest resident butterflies. It is a denizen of
mixed hardwood forests where it frequents woodland
roads or sunlit trails. The relative scarcity of adult sight-
ings is likely due to their preference for spending time
high in the canopy of mature host trees. As the larvae
feed on primarily on developing nuts, stands of young
trees are not utilized. The adults periodically fly down
from the treetops in search of nectar. They also regularly visit moist areas to sip moisture where they may
occasionally be seen in numbers. It typically rests and
feeds with its wings closed.

# Early Hairstreak
*Erora laeta*

**Family/Subfamily:** Gossamer Wings (Lycaenidae)/ Hairstreaks (Theclinae)

**Wingspan:** 0.75–1.00" (1.9–2.5 cm)

**Above:** slate gray with blue scaling toward wing bases; tailless

**Below:** pale grayish green with a band of white-rimmed reddish orange spots across the wings; hindwing has a second row of smaller white-rimmed reddish orange spots along the outer margin

**Sexes:** dissimilar; female has increased iridescent blue scaling above with broad, dark gray borders

**Egg:** pale green laid singly on host leaves, buds, developing fruits and catkins

**Larva:** yellow green to rust brown with large reddish brown patches on the thorax and abdomen

**Larval Host Plants:** American Beech and Beaked Hazelnut

**Habitat:** hardwood forests and clearings, along woodland margins, sun dappled trails, stream corridors and roadsides

**Broods:** two or three generations

**Abundance:** rare to uncommon

**Compare:** unique

Resident

| Jan. | Feb. | Mar. | Apr. | May | June | July | Aug. | Sept. | Oct. | Nov. | Dec. |

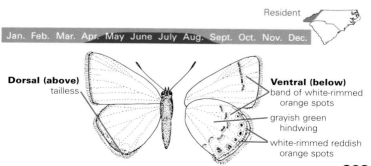

**Dorsal (above)**
tailless

**Ventral (below)**
band of white-rimmed orange spots

grayish green hindwing

white-rimmed reddish orange spots

**263**

Larva

**Comments:** This rare hairstreak is a denizen of Atlantic white cedar swamps and adjacent woodlands. Once common along the outer Coastal Plain, much of this butterfly's freshwater wetland habitat has already disappeared due to extensive logging, fire suppression, hydrologic alteration, and unabated coastal development. As a result, Hessel's Hairstreak now occurs in isolated, highly localized colonies in often widely spaced pockets of remnant habitat.

# Hessel's Hairstreak
*Callophrys hesseli*

**Family/Subfamily:** Gossamer Wings (Lycaenidae)/
Hairstreaks (Theclinae)

**Wingspan:** 0.8–1.1" (2.0–2.8 cm)

**Above:** variable; unmarked dark brown often with some
amber scaling (spring brood individuals are generally
lighter); forewing has pale gray stigma; hindwing has
two short tails

**Below:** blue-green to emerald; forewing has reddish
brown scaling along the trailing margin, a band of
somewhat offset white spots and a single faint white
cell spot; hindwing has two small white bars toward
the base and an irregular white band across the middle, both heavily edged with dark brown

**Sexes:** similar, although female lacks pale forewing
stigma and has increased amber scaling above

**Egg:** light green, laid singly on host

**Larva:** green with white dashes

**Larval Host Plants:** Atlantic White Cedar

**Habitat:** swamps, bogs and adjacent woodlands

**Broods:** two generations

**Abundance:** rare to uncommon

**Compare:** Juniper Hairstreak (pg. 267) is brighter green
below and lacks white forewing cell spot.

Resident

Jan. Feb. Mar. Apr. May June July Aug. Sept. Oct. Nov. Dec.

**Dorsal (above)**

**Ventral (below)**
top white spot set
outwardly

white cell spot

two white bars

white band edged
inwardly in brown

**265**

Larva

**Comments:** This diminutive, reclusive butterfly, although widespread, typically occurs in spotty, localized colonies but can be quite abundant when encountered. It is always found in extremely close association with stands of Eastern Redcedar. This relationship is so intimate that the butterfly spends the majority of its adult life directly on host trees, leaving only occasionally to nectar at nearby blossoms or disperse to pioneer new colonies. The adults regularly perch high in the branches and may go unnoticed. They are best discovered by gently tapping the trunk or branches.

## Juniper Hairstreak
*Callophrys gryneus*

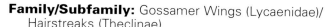

**Family/Subfamily:** Gossamer Wings (Lycaenidae)/ Hairstreaks (Theclinae)

**Wingspan:** 0.8–1.1" (2.0–2.8 cm)

**Above:** variable; unmarked brown to brown with extensive amber scaling; hindwing has two short tails; male has pale gray stigma on forewing

**Below:** bright olive green; forewing has a straight submarginal white band and reddish orange scaling along trailing edge; hindwing has two small white bars toward the base and an irregular white band across the middle that is edged on the inside with reddish brown

**Sexes:** similar, although female lacks pale forewing stigma

**Egg:** light green, laid singly on host

**Larva:** bright green with bold white dashes

**Larval Host Plants:** Eastern Redcedar

**Habitat:** old fields, forest edges, rocky outcrops, coastal areas, dry hillsides, bluffs, rural roadsides, windbreaks, old historical properties and even cemeteries

**Broods:** multiple generations

**Abundance:** uncommon to common

**Compare:** Hessel's Hairstreak (pg. 265) is emerald green with a ventral white forewing cell spot.

Resident

Jan. Feb. Mar. Apr. May June July Aug. Sept. Oct. Nov. Dec.

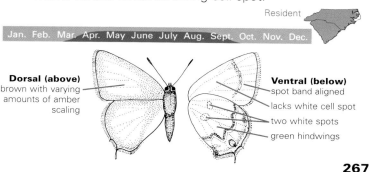

**Dorsal (above)**
brown with varying amounts of amber scaling

**Ventral (below)**
spot band aligned

lacks white cell spot

two white spots

green hindwings

**267**

Larva

**Comments:** Easily overlooked, the Southern Skipperling
is the smallest skipper in the Carolinas. It occasionally
visits home gardens. Adults have a weak, darting flight
and skim quickly over low vegetation stopping now
and then to perch on available leaves or grass blades.
The butterfly may be encountered year-round in
coastal South Carolina and along the Gulf Coast from
Texas to Florida.

# Southern Skipperling
*Copaeodes minima*

**Family/Subfamily:** Skippers (Hesperiidae)/ Banded Skippers (Hesperiinae)

**Wingspan:** 0.50–0.75" (1.3–1.9 cm)

**Above:** elongated wings; male is bright tawny orange; female is brownish orange with darker scaling toward wing bases and along margins

**Below:** yellow orange with narrow, distinctive cream streak through hindwing

**Sexes:** similar, although female somewhat darker

**Egg:** laid singly on host leaves

**Larva:** green

**Larval Host Plants:** grasses including Bermuda Grass

**Habitat:** open, grassy areas including roadsides, utility easements, old fields, meadows and forest edges

**Broods:** multiple generations

**Abundance:** occasional to common

**Compare:** Least Skipper (pg. 273) is larger, has more rounded wings with dark dorsal borders and lacks cream ventral hindwing streak.

Resident Visitor

Jan. Feb. Mar. Apr. May June July Aug. Sept. Oct. Nov. Dec.

**Dorsal (above)** elongated wings

**Ventral (below)** all yellow orange
cream streak

**269**

Ventral

**Comments:** This dainty little butterfly is the only metal-mark in the Carolinas. It typically occurs in spotty, localized colonies in close association with its sole larval host, but may be common when encountered. While not currently in need of conservation, the species appears to be declining throughout its range. Adults scurry low to the ground and frequently stop to perch with their wings held in a characteristically open posture. If disturbed, or during inclement weather, they dart away and land out of sight on the underside of large, broad leaves.

# Little Metalmark
*Calephelis virginiensis*

**Family/Subfamily:** Gossamer Wings (Lycaenidae)/ Metalmarks (Riodininae)

**Wingspan:** 0.5–1.0" (1.3–2.5 cm)

**Above:** brownish orange with numerous delicate dark markings and two narrow, metallic gray bands along the outer edge of the wings

**Below:** similar to upper surface, but paler; brownish orange with dark spots and metallic gray bands

**Sexes:** similar, although female has broader, more rounded wings

**Egg:** laid singly on host leaves

**Larva:** pale green with long white hairs

**Larval Host Plants:** Yellow Thistle

**Habitat:** open grassy areas, pine savannas, moist meadows and forest edges

**Broods:** multiple generations

**Abundance:** occasional, locally common

**Compare:** unique

Resident

Jan. Feb. Mar. Apr. May June July Aug. Sept. Oct. Nov. Dec.

**Dorsal (above)**
brownish orange wings

dark reticulations

metallic bands

**Ventral (below)**
metallic bands

**Comments:** The Least Skipper has a low, weak flight and flutters slowly through the vegetation, occasionally pausing to nectar or perch. They regularly visit flowers but prefer low plants with small blossoms. It is not a regular garden visitor. Adults have noticeably rounded wings. Although the sexes are similar, the male has a very long slender abdomen that protrudes well past the wings.

# Least Skipper
*Ancyloxypha numitor*

**Family/Subfamily:** Skippers (Hesperiidae)/
Banded Skippers (Hesperiinae)

**Wingspan:** 0.7–1.0" (1.8–2.5 cm)

**Above:** forewing orange-brown with dark border; hind-wing orange with dark border; rounded wings; male has a long, pointed abdomen

**Below:** forewing dark brown with orange border; hind-wing orange-gold

**Sexes:** similar, although female has a shorter abdomen

**Egg:** yellow, laid singly on or near host leaves

**Larva:** long and slender, light yellow-green with thin dark dorsal stripe and reddish brown head; head has numerous cream stripes

**Larval Host Plants:** various grasses including Rice Cutgrass, Giant Cutgrass, panic grass and cordgrass

**Habitat:** moist, grassy areas including roadside ditches, utility easements, wet meadows, pond edges and old fields

**Broods:** multiple generations

**Abundance:** occasional to common

**Compare:** The Southern Skipperling (pg. 269) is smaller, has elongated wings and a distinct pale ray through the hindwing below.

Resident

Jan. Feb. Mar. Apr. May June July Aug. Sept. Oct. Nov. Dec.

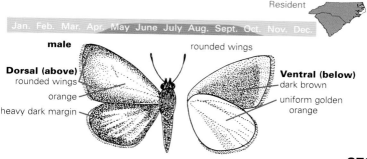

**male**

rounded wings

**Dorsal (above)**
rounded wings
orange
heavy dark margin

**Ventral (below)**
dark brown
uniform golden orange

**273**

Ventral

Larva

**Comments:** This common skipper of the northeast just barely enters our area in the western mountains of North Carolina. As its name implies, the species was accidentally introduced into Ontario, Canada from Europe in 1910 and continues to expand its range. Superficially similar to the Least Skipper, the butterfly is more likely to be found in dryer areas of open grassy habitat. Univoltine (only one brood per year) over its entire range, the tiny white eggs overwinter and hatch the following spring.

# European Skipper
*Thymelicus lineola*

**Family/Subfamily:** Skippers (Hesperiidae)/ Banded Skippers (Hesperiinae)

**Wingspan:** 0.9–1.1" (2.3–2.8 cm)

**Above:** bronzy orange wings with dark brown borders; veins are darkened toward the outer wing margins

**Below:** unmarked orange

**Sexes:** similar, although female is somewhat darker with veins darkened to base

**Egg:** white, laid on host stems

**Larva:** pale green with a darker green dorsal stripe, white lateral stripes and a greenish tan head marked with two vertical cream stripes on the face

**Larval Host Plants:** various grasses including Timothy Grass, Orchardgrass, Common Velvetgrass and bent-grass

**Habitat:** open, grassy areas including roadsides, utility easements, old fields, meadows and pastures

**Broods:** single generation

**Abundance:** occasional

**Compare:** Least Skipper (pg. 273) is more widespread in the Carolinas, has more rounded wings, solid dark dorsal borders, and multiple generations.

Resident

Jan. Feb. Mar. Apr. May June July Aug. Sept. Oct. Nov. Dec.

**Dorsal (above)**
dark brown borders

veins darker toward outer margins

bronzy orange

**Ventral (below)**
unmarked orange

**275**

Male

Male

**Comments:** The Phaon Crescent is a temporary colonist of southern portions of the Carolinas, and its population is spotty and variable from year to year. It remains close to its low-growing host, but may be an occasional garden visitor. Adults have a low, rapid flight and are easily disturbed. Males perch on low vegetation and frequently patrol for females. They occasionally gather at moist ground. The larvae are gregarious when young but become solitary towards maturity.

# Phaon Crescent
*Phyciodes phaon*

**Family/Subfamily:** Brush-foots (Nymphalidae)/
True Brush-foots (Nymphalinae)

**Wingspan:** 0.90–1.25" (2.3–3.2 cm)

**Above:** orange with dark spots, bands and wing borders;
forewing has a distinctive pale yellow to cream postmedian band

**Below:** seasonally variable; cream with brown bands,
spots and patches; winter-form has increased brown
coloration on ventral hindwings

**Sexes:** similar

**Egg:** light green, laid in clusters on the underside of host
leaves

**Larva:** amber brown with dark brown lines and short,
branched spines

**Larval Host Plants:** Frogfruit

**Habitat:** open, disturbed sites including roadsides, old
fields, moist ditches, utility easements, fallow agricultural land and pond edges

**Broods:** multiple generations

**Abundance:** occasional to common; locally abundant

**Compare:** Pearl Crescent (pg. 301) is tawny orange
below and lacks dorsal pale yellow median forewing
band.

Visitor

Jan. Feb. Mar. Apr. May June July Aug. Sept. Oct. Nov. Dec.

**Dorsal (above)**
pale median band

extensive black
markings and
borders

**Ventral (below)**
seasonally variable

tan with fine brown
lines

pale crescent in dark
marginal patch

**Comments:** The Fiery Skipper has a rapid, darting flight but frequently stops to perch on low vegetation. Adults are exceedingly fond of flowers and readily congregate at available blossoms. They have a strong preference for colorful composites. The larvae utilize a variety of grasses including many commonly planted for lawns. As a result, the butterfly is often mentioned as a minor turf pest. It is often found alongside the similar-looking and equally abundant Sachem and Whirlabout.

# Fiery Skipper
*Hylephila phyleus*

**Family/Subfamily:** Skippers (Hesperiidae)/
Banded Skippers (Hesperiinae)

**Wingspan:** 1.00–1.25" (2.5–3.2 cm)

**Above:** elongated wings; male is golden orange with
jagged black border and black stigma; female is tawny
orange with dark brown bands

**Below:** hindwing golden orange in male or orange brown
in female with tiny dark brown spots

**Sexes:** dissimilar; female darker with reduced orange
markings and larger hindwing spots

**Egg:** whitish green, laid singly on host leaves

**Larva:** greenish brown with thin, dark brown dorsal
stripe and black head

**Larval Host Plants:** a variety of grasses including
Bermuda Grass, crabgrass, bentgrass and St.
Augustine Grass

**Habitat:** open, grassy areas including old fields, roadsides, vacant lots, open woodlands, forest edges,
parks, lawns and gardens

**Broods:** multiple generations

**Abundance:** common to abundant

**Compare:** Whirlabout (pg. 281) has two rows of large
dark spots on the ventral hindwing. Sachem (pg. 293)
lacks dark spots on ventral hindwing.

Visitor

Jan. Feb. Mar. Apr. May June July Aug. Sept. Oct. Nov. Dec.

male

**Dorsal (above)**
orange

jagged black
margins

**Ventral (below)**
small scattered dark
spots

**279**

Male

Female pg. 135

**Comments:** The Whirlabout is a diminutive skipper with two distinct rows of squarish spots on the hindwings below. This species is sexually dimorphic. Males are tawny orange and considerably brighter than their drab brown female counterparts. It regularly expands its range each summer, establishing temporary breeding colonies throughout the southeast from Maryland to Texas. Living up to its name, adults have a low, erratic flight and scurry quickly around, periodically stopping to perch or nectar. Avidly fond of flowers, the butterfly is a frequent garden visitor.

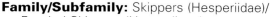

## Whirlabout
*Polites vibex*

**Family/Subfamily:** Skippers (Hesperiidae)/
Banded Skippers (Hesperiinae)

**Wingspan:** 1.00–1.25" (2.5–3.2 cm)

**Above:** elongated wings; golden orange with black borders and black stigma; female is dark brown with cream spots on forewing

**Below:** hindwing yellow in male or bronze brown in female with two loose bands of large dark brown spots

**Sexes:** dissimilar; female brown above with little orange scaling; olive brown below with similar pattern as male

**Egg:** white, laid singly on host leaves

**Larva:** brownish green with thin, dark dorsal stripe and black head

**Larval Host Plants:** various grasses including Bermuda Grass, St. Augustine Grass and Crabgrass

**Habitat:** open, disturbed areas including old fields, roadsides, vacant lots, open woodlands, forest edges, parks, lawns and gardens

**Broods:** multiple generations

**Abundance:** occasional to abundant

**Compare:** Fiery Skipper (pg. 279) has more elongated forewings and scattered, tiny dark spots on ventral hindwing.

Resident Visitor

Jan. Feb. Mar. Apr. May June July Aug. Sept. Oct. Nov. Dec.

**male**

**Dorsal (above)**
golden orange
large black stigma
jagged black border
orange
smooth black border

**Ventral (below)**
two bands of large dark brown spots
golden orange

**281**

Ventral

Larva

**Comments:** Although called the American Copper, some suggest that the eastern populations of the butterfly may actually be the result of historical introductions from Europe. This argument is fueled by the species' unique preference for open, disturbed habitats and primary use of a weedy, non-native larval host. It tends to occur in widespread and highly localized colonies, but is often fairly common when encountered. The adults frequently perch on bare soil or on low vegetation with the wings held in a characteristic, partially open posture. When first seen in bright sunlight, there remains little doubt as to why the butterfly is called a copper!

# American Copper
*Lycaena phlaeas*

**Family/Subfamily:** Gossamer Wings (Lycaenidae)/ Coppers (Lycaeninae)

**Wingspan:** 0.9–1.4" (2.3–3.6 cm)

**Above:** bright red orange forewings with black spots and dark borders; hindwing is gray with wide scalloped submarginal orange band subtended by black spots along outer margin

**Below:** forewing is pale orange with prominent white-rimmed black spots and light gray apex and outer margin; hindwing is silvery gray with small white-rimmed black spots and narrow, irregular reddish orange line along outer margin

**Sexes:** similar, although female is larger and has more rounded wings

**Egg:** pale greenish white, laid singly host stems or leaves

**Larva:** variable; yellow green to rose, often with a narrow lateral stripe

**Larval Host Plants:** Sheep Sorrel and Curly Dock

**Habitat:** open, disturbed sites, old fields, utility easements, roadsides, pastures and meadows

**Broods:** two or more generations

**Abundance:** occasional

**Compare:** unique

Resident

Jan. Feb. Mar. Apr. May June July Aug. Sept. Oct. Nov. Dec.

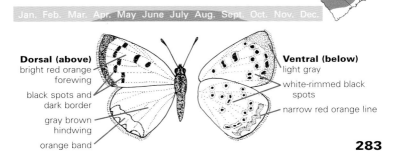

**Dorsal (above)**
bright red orange forewing

black spots and dark border

gray brown hindwing

orange band

**Ventral (below)**
light gray

white-rimmed black spots

narrow red orange line

**283**

**Comments:** Unique in appearance, the Harvester is the only North American butterfly with predaceous larvae. It tends to be very localized and is seldom encountered in large numbers. Males regularly perch on sunlit leaves with their wings partially open. The Harvester has a fast, erratic flight reminiscent of most hairstreaks and can be a challenge to follow. Adults feed primarily on aphid honeydew and rarely visit flowers.

## Harvester
*Feniseca tarquinius*

**Family/Subfamily:** Gossamer Wings (Lycaenidae)/ Harvesters (Miletinae)

**Wingspan:** 1.1–1.3" (2.8–3.3 cm)

**Above:** orange with brown to black spots, patches and borders

**Below:** brown; forewing has orange central scaling and several dark patches outlined in white; hindwing has numerous dark spots outlined in white and silver scaling toward base

**Sexes:** similar

**Egg:** greenish white, laid singly among aphid colonies

**Larva:** gray with whitish yellow bumps bordered with brown along top, reddish brown lateral stripes and long gray hairs

**Larval Host Plants:** carnivorous on woolly aphids

**Habitat:** forest edges, moist woodlands and associated clearings, trails, waterways and roads

**Broods:** multiple generations

**Abundance:** occasional

**Compare:** unique

Resident

Jan. Feb. Mar. Apr. May June July Aug. Sept. Oct. Nov. Dec.

**female**

**Dorsal (above)**
orange with black border and spots

**Ventral (below)**
reddish brown with silver scaling and numerous brown spots outlined in white

Female

Ventral

**Comments:** The Delaware Skipper is a small orange butterfly with elongated, somewhat pointed forewings.
The ventral wing surface is immaculate golden yellow.
Found primarily in a variety of moist, grassy habitats
from damp meadows to coastal marshes, the butterfly
frequently finds its way into suburban yards. Adults
have a quick, darting flight and are fond of flowers.
Males perch on low leaves and grasses and make frequent exploratory flights.

# Delaware Skipper
*Anatrytone logan*

**Family/Subfamily:** Skippers (Hesperiidae)/
Banded Skippers (Hesperiinae)

**Wingspan:** 1.0–1.4" (2.5–3.6 cm)

**Above:** orange with dark borders and veins; forewings
are elongated and somewhat pointed; male has a
small black cell-end bar on forewing; female has
brown scaling in forewing cell, wider borders, and
forewing cell-end bar is larger

**Below:** unmarked golden orange

**Sexes:** similar, although female darker with reduced
orange coloration

**Egg:** white, laid singly on host leaves

**Larva:** bluish white with dark tubercles and a black-and-
white head

**Larval Host Plants:** various grasses including
bluestems, silver plumegrass and Switch Grass

**Habitat:** open woodlands, forest edges, roadsides, pastures, wetland edges and old fields

**Broods:** two generations

**Abundance:** occasional to common

**Compare:** Byssus Skipper (pg. 317) has more pointed
forewings and pale yellow spots on the ventral hindwing. European Skipper (pg. 275) lacks black forewing
cell-end bar.

Resident

Jan. Feb. Mar. Apr. May June July Aug. Sept. Oct. Nov. Dec.

**male**

**Dorsal (above)**
dark border
thin black cell-end bar
dark veins

**Ventral (below)**
unmarked golden
orange

**287**

Male

Female pg. 147    Male    Female

**Comments:** Sexually dimorphic, the Zabulon Skipper is a small butterfly with a bright orange male and a drab brown female. Primarily a butterfly of woodlands and adjacent open sites, it occasionally wanders into sub-urban gardens. Adults have a rapid flight and are strongly attracted to flowers. Males perch on sunlit branches along trails or clearings and engage passing objects or rival males. Females generally prefer to remain within the confines of nearby shady sites.

# Zabulon Skipper
*Poanes zabulon*

**Family/Subfamily:** Skippers (Hesperiidae)/
Banded Skippers (Hesperiinae)

**Wingspan:** 1.0–1.4" (2.5–3.6 cm)

**Above:** male is golden orange with dark brown borders
and small brown spot near forewing apex; female is
dark brown with band of cream spots across forewing

**Below:** male hindwing yellow with a brown base enclosing a yellow spot; female is dark brown with small
light subapical spots, lavender scaling on wing margins, and white bar along leading margin of hindwing

**Sexes:** dissimilar, female brown with little orange color

**Egg:** pale green, laid singly on host leaves

**Larva:** tan with dark dorsal stripe, white lateral stripe and
short, light-colored hairs; reddish brown head

**Larval Host Plants:** various grasses including
Purpletop Grass, Whitegrass and lovegrass

**Habitat:** open woodlands, forest edges, roadsides, pastures, wetland edges, stream corridors and old fields

**Broods:** two generations

**Abundance:** occasional

**Compare:** Sachem (pg. 293) is similar to male Zabulon
Skipper, but dark hindwing base does not enclose yellow patch.

Resident

Jan. Feb. Mar. Apr. May June July Aug. Sept. Oct. Nov. Dec.

male

Dorsal (above)
dark spot

narrow black cell
end bar

golden orange

clear golden
orange

Ventral (below)
dark base encloses
yellow spot

yellow with darker
spots

**289**

Ventral

Larva

**Comments:** The small, ornately patterned Gorgone
Checkerspot reaches the easternmost extension of its
large range just across the western border of the
Carolinas, where it is a localized and rare resident. It
frequents open deciduous woodlands and adjacent disturbed areas including utility easements and logged
sites. Adults have a low, erratic flight and frequently
stop to nectar at available flowers where their distinctive markings can be closely observed. An excellent
colonizer, adults frequently wander far to establish
new populations. Young larvae feed gregariously. Late
instar larvae overwinter.

# Gorgone Checkerspot
*Chlosyne gorgone*

**Family/Subfamily:** Brush-foots (Nymphalidae)/
True Brush-foots (Nymphalinae)

**Wingspan:** 1.1–1.4" (2.8–3.6 cm)

**Above:** tawny orange with black bands and spots

**Below:** hindwing has a distinct zigzag pattern of white
chevrons between darker bands

**Sexes:** similar

**Egg:** pale green, laid in clusters on host leaves

**Larva:** yellow orange with black longitudinal stripes,
black branching spines and a black head

**Larval Host Plants:** various composites including Great
Ragweed and sunflowers

**Habitat:** open woodlands, stream corridors, forest
edges, and adjacent dry, open, grassy areas, utility
easements and previously burned sites

**Broods:** multiple generations

**Abundance:** rare to occasional

**Compare:** Silvery Checkerspot (pg. 323) is larger, its dorsal hindwing has white-centered black spots and its
ventral hindwing has a dark marginal patch, an incomplete row of silvery white crescents along the margin,
and lacks zig-zag pattern.

Resident

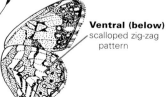

Jan. Feb. Mar. Apr. May June July Aug. Sept. Oct. Nov. Dec.

**Dorsal (above)**
orange with black
bands and spots

pale orange chevrons
in black border

**Ventral (below)**
scalloped zig-zag
pattern

**291**

Male

Female pg. 155    Male    Female

**Comments:** The Sachem shares its affinity for open, disturbed sites with the Whirlabout and Fiery Skipper
with which it often flies. Adults have a quick, darting
flight that is usually low to the ground. All three
species are somewhat nervous butterflies, but readily
congregate at available flowers. Together, they often
form a circus of activity with several pausing briefly to
perch or nectar before one flies up and disturbs the
others, only to alight again moments later.

# Sachem
*Atalopedes campestris*

**Family/Subfamily:** Skippers (Hesperiidae)/
Banded Skippers (Hesperiinae)

**Wingspan:** 1.0–1.5" (2.5–3.8 cm)

**Above:** elongated wings; male is golden orange with
brown borders and large, black stigma; female is dark
brown with golden markings in wing centers; forewing
has black median spot and several semitransparent
spots

**Below:** variable; hindwing golden brown in male, brown
in female with pale postmedian patch or band of spots

**Sexes:** dissimilar; female darker with semitransparent
forewing spots and reduced orange markings

**Egg:** white, laid singly on host leaves

**Larva:** greenish brown with thin, dark dorsal stripe and
black head

**Larval Host Plants:** various grasses including Bermuda
Grass and Crabgrass

**Habitat:** open, disturbed areas including old fields, pastures, roadsides, parks, lawns and gardens

**Broods:** multiple generations

**Abundance:** common to abundant

**Compare:** Fiery Skipper (pg. 279) and Whirlabout (pg.
281) have dark spots on the ventral hindwing.

Resident Visitor

Jan. Feb. Mar. Apr. May June July Aug. Sept. Oct. Nov. Dec.

**male**

**Dorsal (above)**
large rectangular
stigma
golden orange

**Ventral (below)**
faint dark spot along
trailing margin
large pale patch

Ventral

Larva

**Comments:** The Arogos Skipper is a rare and declining species throughout most of the Southeast. Limited by habitat, it typically occurs in isolated, localized colonies in close proximity to undisturbed stands of its larval hosts. The larvae construct individual tent-like shelters on the host by weaving together two leaves with silk. They overwinter partially grown and complete development the following spring. Additional research and status surveys are badly needed to update the current distribution of this butterfly in the Carolinas.

# Arogos Skipper
*Atrytone arogos*

**Family/Subfamily:** Skippers (Hesperiidae)/
Banded Skippers (Hesperiinae)

**Wingspan:** 1.2–1.4" (3.0–3.6 cm)

**Above:** clear yellow orange with smooth, broad, dark
brown borders and white fringes; forewings are
pointed

**Below:** unmarked golden orange, occasionally with
lighter veins

**Sexes:** similar, although female has broader dark borders
and reduced orange scaling above

**Egg:** pale yellow, laid singly on host leaves

**Larva:** pale green with a dark green dorsal line and a
gray white head marked with orange brown vertical
lines

**Larval Host Plants:** various grasses including Big
Bluestem and Little Bluestem

**Habitat:** pine flatwoods, sandhills and barrens

**Broods:** two generations

**Abundance:** rare

**Compare:** Delaware Skipper (pg. 287) is much more
widespread and common, has dark veins on the
forewings above and is brighter orange below.

Resident

Jan. Feb. Mar. Apr. May June July Aug. Sept. Oct. Nov. Dec.

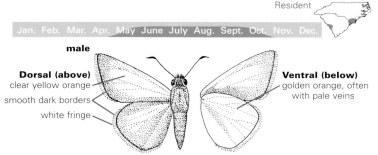

**male**

**Dorsal (above)**
clear yellow orange
smooth dark borders
white fringe

**Ventral (below)**
golden orange, often
with pale veins

**295**

Ventral

Female

Larva

**Comments:** Although widespread throughout the
Northeast, this lovely tawny orange skipper has an
extremely restricted range within our area, being limited solely to the western mountains of North Carolina.
Producing only one generation each year, the adults
are on the wing in late spring and early summer.
Spotty and localized, colonies may be easily overlooked especially due to the brown, early-season
appearance of its preferred habitats. Both sexes readily visit a variety of spring wildflowers. Larvae
construct individual leaf shelters at the base of the
host.

# Indian Skipper
*Hesperia sassacus*

**Family/Subfamily:** Skippers (Hesperiidae)/
Banded Skippers (Hesperiinae)

**Wingspan:** 1.2–1.4" (3.0–3.6 cm)

**Above:** bright orange with wide, somewhat jagged, dark brown borders

**Below:** hindwing is tawny orange with a pale golden orange spot band through center; middle spot displaced outward toward margin

**Sexes:** similar, although female has broader dark borders, larger, paler yellow orange spot bands and reduced orange scaling

**Egg:** whitish green, laid singly on host leaves or stems

**Larva:** dark brown, often with lighter mottling and a round black head

**Larval Host Plants:** various grasses including Little Bluestem, panic grasses, Poverty Oatgrass, bluegrass and Red Fescue

**Habitat:** woodland clearings and margins, pastures, old brushy fields and forest meadows

**Broods:** single generation

**Abundance:** rare

**Compare:** Female Sachem (pg. 155) has longer, more pointed forewings and glassy forewing spots.

Resident

Jan. Feb. Mar. Apr. May June July Aug. Sept. Oct. Nov. Dec.

**male**

**Dorsal (above)**
dark, sharply
defined borders

**Ventral (below)**
pale spot band with
center spot offset
toward outer margin

Male

Ventral

Larva

**Comments:** This fairly large golden orange skipper is found throughout the Southeast from Texas to North Carolina. Rare and reclusive, it is typically found in spotty and highly localized colonies. Like most skippers, it has a quick darting flight and is difficult to observe unless perched or nectaring. I attracted this beautiful skipper to my garden in southeastern SC on several occasions! Due to its elusive nature, information about this butterfly's life history, behavior and ecology remains poorly known.

# Meske's Skipper
*Hesperia meskei*

**Family/Subfamily:** Skippers (Hesperiidae)/
Banded Skippers (Hesperiinae)

**Wingspan:** 1.2–1.6" (3.0–4.1 cm)

**Above:** dark brown with tawny orange scaling toward
bases and orange spot bands

**Below:** hindwing is bright golden orange with a pale spot
band, often very faint

**Sexes:** similar, although female is darker with broader
wings, large, pale yellow orange spot bands and
reduced basal orange scaling

**Egg:** currently undocumented

**Larva:** green brown; round black head

**Larval Host Plants:** various grasses including
Arrowfeather Threeawn and Little Bluestem

**Habitat:** open woodlands, forest clearings and margins,
and adjacent open areas

**Broods:** two generations

**Abundance:** rare

**Compare:** Indian Skipper (pg. 297) is smaller with more
extensive orange scaling on the wings above, and
more well-defined ventral hindwing spot band.

Resident

Jan. Feb. Mar. Apr. May June July Aug. Sept. Oct. Nov. Dec.

**male**

**Dorsal (above)**
orange spot bands
tawny toward base

**Ventral (below)**
hindwing yellow
orange
faint spot band

**299**

Male

**Comments:** The lovely tawny-orange Pearl Crescent is our most widespread and abundant crescent. It is seasonally variable, and spring and fall individuals are darker and more heavily patterned on the ventral hindwings. It is an opportunistic breeder, continually producing new generations as long as favorable conditions allow. It has a rapid, erratic flight. Males perch on low vegetation with wings outstretched and frequently patrol for females. Freshly emerged males often gather at moist ground.

# Pearl Crescent
*Phyciodes tharos*

**Family/Subfamily:** Brush-foots (Nymphalidae)/
True Brush-foots (Nymphalinae)

**Wingspan:** 1.25–1.60" (3.2–4.1 cm)

**Above:** orange with dark bands, spots and wing borders

**Below:** seasonally variable; light brownish orange with
brown markings; winter-form has increased dark coloration and pattern elements

**Sexes:** similar, although female is paler orange with
increased black markings

**Egg:** green, laid in clusters on the underside of host
leaves

**Larva:** dark brown to charcoal with lateral cream stripes
and numerous short, branched spines

**Larval Host Plants:** various asters including Frost
Aster, Smooth Blue Aster and Bushy Aster

**Habitat:** open, sunny locations including roadsides, old
fields, utility easements, forest edges and gardens

**Broods:** multiple generations

**Abundance:** occasional to common

**Compare:** Phaon Crescent (pg. 277) has a pale yellow
median forewing band and lighter ventral hindwing coloration.

Resident

Jan. Feb. Mar. Apr. May June July Aug. Sept. Oct. Nov. Dec.

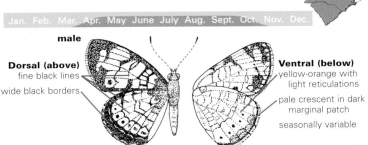

**male**

**Dorsal (above)**
fine black lines
wide black borders

**Ventral (below)**
yellow-orange with
light reticulations

pale crescent in dark
marginal patch

seasonally variable

**301**

Larva

**Comments:** This southeastern skipper is confined to the Atlantic seaboard and Gulf Coast from New Jersey to Texas. Rare and extremely localized, it is an inhabitant of coastal saltwater marshes but may occasionally be found in inland wetlands. Adults visit a variety of moisture-loving flowers. It remains one of our most poorly known and studied butterflies.

# Aaron's Skipper
*Poanes aaroni*

**Family/Subfamily:** Skippers (Hesperiidae)/
Banded Skippers (Hesperiinae)

**Wingspan:** 1.3–1.7" (3.3–4.3 cm)

**Above:** clear golden orange with dark brown borders and
a thin, faint black stigma

**Below:** hindwing is dull orange brown with a narrow pale
ray in the center

**Sexes:** dissimilar; female is darker with more rounded
wings and has reduced golden orange scaling

**Egg:** white, laid singly on host leaves

**Larva:** pale brown with dark longitudinal stripes; reddish
brown head

**Larval Host Plants:** in association with Smooth
Cordgrass

**Habitat:** coastal salt marshes, some inland freshwater
marshes

**Broods:** two generations

**Abundance:** rare to occasional

**Compare:** Broad-winged Skipper (pg. 327) is larger and
has a broader, more distinct pale hindwing ray below.

Resident

Jan. Feb. Mar. Apr. May June July Aug. Sept. Oct. Nov. Dec.

male

**Dorsal (above)**
clear golden orange
dark borders

**Ventral (below)**
pale ray
dull orange brown

**303**

Male

Female    Ventral    Female    Larva
                     "Pocahontas"
                     pg. 195

**Comments:** This small woodland skipper has a single
early summer flight. Males perch on sunlit leaves and
aggressively dart out at other passing butterflies.
Although common throughout the north and east, its
range barely enters the extreme western mountains of
the Carolinas. The female produces two distinct forms.
The lighter form closely resembles the male and the
darker form "Pocahontas" is superficially similar to
Zabulon Skipper females. Larvae construct individual
leaf shelters on the host. Larvae overwinter and complete development the following spring.

# Hobomok Skipper
*Poanes hobomok*

**Family/Subfamily:** Skippers (Hesperiidae)/
Banded Skippers (Hesperiinae)

**Wingspan:** 1.4–1.6" (3.6–4.1 cm)

**Above:** golden orange with irregular dark brown borders
and a narrow black cell-end bar on the forewing

**Below:** purplish brown with a broad yellow orange patch
through the hindwing

**Sexes:** dissimilar; female has two forms. Normal form
resembles male but has reduced orange scaling above.
"Pocahontas" form is dark brown above with pale
forewing spots; hindwing is purplish brown below with
faint band and violet gray frosting along outer margin.

**Egg:** white, laid singly on host leaves

**Larva:** brown green with numerous short, light-colored
hairs; round, brown head

**Larval Host Plants:** various grasses including Little
Bluestem, panic grasses, Poverty Oatgrass, bluegrass
and Rice Cutgrass

**Habitat:** open woodlands, forest edges, clearings and
trails, roadsides and along forested stream margins

**Broods:** single generation

**Abundance:** occasional to common

**Compare:** Zabulon (pg. 289) and Peck's (pg. 133)
Skippers have yellow basal scaling on ventral hindwing.

Resident

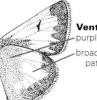

Jan. Feb. Mar. Apr. May June July Aug. Sept. Oct. Nov. Dec.

**male**

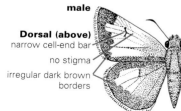

**Dorsal (above)**
narrow cell-end bar
no stigma
irregular dark brown
borders

**Ventral (below)**
purplish brown
broad yellow orange
patch

**305**

Larva

**Comments:** The Tawny Crescent is a rare resident
species that has severely declined in recent years over
much of its eastern range. Highly restricted within the
Carolinas, the species is of conservation concern and
strongly in need of status surveys to determine the
location and number of remaining colonies. Due to is
close resemblance to other crescents, it is possible
that many localized populations have been overlooked.
Adults have a quick, somewhat erratic flight and frequently alight on low vegetation with their wings open.
Males occasionally visit mud puddles to sip moisture.
Later instar larvae overwinter.

# Tawny Crescent
*Phyciodes batesii*

**Family/Subfamily:** Brush-foots (Nymphalidae)/
True Brush-foots (Nymphalinae)

**Wingspan:** 1.3–1.75" (3.3–4.3 cm)

**Above:** tawny orange with heavy black bands, spots and
broad wing borders

**Below:** hindwing is golden yellow brown with very fine
brown markings and a postmedian row of small dark
spots

**Sexes:** similar

**Egg:** pale green, laid in clusters on the underside of host
leaves

**Larva:** pinkish brown with a dark dorsal band, a lateral
yellow stripe and numerous short, light brown spines

**Larval Host Plants:** various asters including Waxyleaf
Aster

**Habitat:** woodland openings, pastures and hillsides

**Broods:** single generation

**Abundance:** rare to uncommon

**Compare:** Pearl Crescent (pg. 301) is has more extensive open orange areas on wings above and has a
distinct dark marginal patch on the ventral hindwing.

Resident

Jan. Feb. Mar. Apr. May June July Aug. Sept. Oct. Nov. Dec.

**Dorsal (above)**

**Ventral (below)**
lacks dark patch
below apex

broad, straight dark
band

fine brown markings

row of small dark
spots

**307**

Male

Female pg. 199

Ventral

Larva

**Comments:** The Yehl Skipper is rare to uncommon large butterfly of southeastern forested swamps or moist woodlands. It is most frequently encountered in sunlit clearings, margins or roadsides. The adults are very wary and often difficult to closely approach. They visit a variety of moisture-loving flowers. Little detailed information is known about the biology and behavior of this reclusive species.

# Yehl Skipper
*Poanes yehl*

**Family/Subfamily:** Skippers (Hesperiidae)/
Banded Skippers (Hesperiinae)

**Wingspan:** 1.30–1.75" (3.3–4.3 cm)

**Above:** golden orange with broad dark brown borders;
forewing has long, narrow black stigma; female is
overall brownish

**Below:** hindwing is yellow orange with a pale band of
spots (usually three prominent spots)

**Sexes:** dissimilar; female has pale forewing spots and
reduced golden orange scaling on the wing bases
above; ventral hindwing is brown with pale spot band

**Egg:** currently undocumented

**Larva:** green brown with a dark dorsal stripe; dark brown
head

**Larval Host Plants:** currently undocumented

**Habitat:** forested swamps and associated clearings and
margins

**Broods:** two generations

**Abundance:** rare to occasional to abundant

**Compare:** Broad-winged Skipper (pg. 327) has broad
pale ray through the spot band on the hindwing below.

Resident

Jan. Feb. Mar. Apr. May June July Aug. Sept. Oct. Nov. Dec.

**male**

**Dorsal (above)**
long, narrow black
stigma

broad, dark brown
borders

**Ventral (below)**
three to four pale
spots in a band

Ventral

Male

Female

Larva

**Comments:** Another large skipper of open swamps and marshes, this lovely tawny orange species has a distinctive pale ray through the ventral hindwing. Although widespread geographically throughout the East, the Dion Skipper is generally quite localized and uncommon. Adults are somewhat sluggish fliers through wetland vegetation but can move very quickly if disturbed. Males readily perch on the tops of sedges or grasses but tend to be wary and difficult to closely approach. They do not wander far from their wetland haunts. Larvae overwinter.

# Dion Skipper
*Euphyes dion*

**Family/Subfamily:** Skippers (Hesperiidae)/
Banded Skippers (Hesperiinae)

**Wingspan:** 1.4–1.7" (3.6–4.3 cm)

**Above:** forewing is orange with broad, dark brown borders and a prominent black forewing stigma; hindwing
is faint orange with a dark brown border and distinctly
brighter elongated orange spot

**Below:** hindwing is tawny orange with faint light ray
through the center

**Sexes:** dissimilar; female is primarily dark brown dorsally
with yellow orange forewing spots and prominent single elongated orange hindwing spot

**Egg:** light green, laid singly on host leaves

**Larva:** blue-green with darker green dorsal line; white
head marked with a black forehead spot bordered by
orange brown vertical lines

**Larval Host Plants:** various sedges including
Woolgrass and Shoreline Sedge

**Habitat:** open wetlands, marshes and swamps

**Broods:** two generations

**Abundance:** rare to occasional

**Compare:** Duke's Skipper (pg. 319) lacks bright orange
scaling above, has more rounded wings and prefers
more shaded habitats.

Resident

Jan. Feb. Mar. Apr. May June July Aug. Sept. Oct. Nov. Dec.

male

**Dorsal (above)**
prominent stigma
broad, dark borders
elongated orange
spot

**Ventral (below)**
pale ray
tawny orange to
reddish brown

**311**

Ventral

Larva

**Comments:** Our smallest fritillary, this butterfly is restricted to extreme northwestern portions of the Carolinas. As its name suggests, the Meadow Fritillary inhabits a variety of open, sunny landscapes from moist pastures to old fields and is quite tolerant of disturbed sites. Adults have a rapid and somewhat erratic flight, typically low to the ground just above the vegetation, but frequently stop to nectar at available wildflowers. Later stage larvae overwinter.

## Meadow Fritillary
*Boloria bellona*

**Family/Subfamily:** Brush-foots (Nymphalidae)/ Longwing Butterflies (Heliconiinae)

**Wingspan:** 1.25–1.90" (3.2–4.8 cm)

**Above:** orange with black bands and spots; elongated wings with squared-off forewing apex

**Below:** hindwing is mottled brown orange with violet frosting along the hindwing margin

**Sexes:** similar

**Egg:** tiny cream eggs laid singly and somewhat haphazardly near host

**Larva:** purplish black with fine yellow mottling and short cream-based brown spines

**Larval Host Plants:** violets

**Habitat:** old fields, meadows, pastures and roadsides

**Broods:** two generations

**Abundance:** occasional

**Compare:** unique

Resident

Jan. Feb. Mar. Apr. May June July Aug. Sept. Oct. Nov. Dec.

**Dorsal (above)**
squared-off apex
basal half of wings darker

**Ventral (below)**
mottled brown orange
violet frosting along margin

Larva

**Comments:** Berry's Skipper is easily one of our rarest
butterflies. Limited entirely to the Southeast from
southern South Carolina to the Florida panhandle, it is
restricted to wetlands and their associated margins.
Occurring in spotty and highly localized colonies, it is
seldom encountered in large numbers. Due to its
reclusive nature, little is known about the species' biology or behavior.

# Berry's Skipper
*Euphyes beryi*

**Family/Subfamily:** Skippers (Hesperiidae)/
Banded Skippers (Hesperiinae)

**Wingspan:** 1.4–1.8" (3.6–4.6 cm)

**Above:** orange with broad, dark brown borders and a
prominent black stigma

**Below:** hindwing is tawny orange with light veins

**Sexes:** dissimilar; female is primarily dark brown with a
few yellow orange spots, including one in the forewing
cell

**Egg:** currently undocumented

**Larva:** pale green with a dark dorsal stripe; reddish
brown head bears a black spot on forehead surrounded by white

**Larval Host Plants:** various sedges

**Habitat:** swamps, wetland margins including pond
edges and along canals

**Broods:** two generations

**Abundance:** very rare and localized

**Compare:** Dion Skipper (pg. 311) has pale ray through
the hindwing below.

Resident

Jan. Feb. Mar. Apr. May June July Aug. Sept. Oct. Nov. Dec.

**male**

**Dorsal (above)**
broad dark borders

**Ventral (below)**
orange brown with
light veins

Male  Female  Larva

**Comments:** The Byssus Skipper is a butterfly of salt and
brackish water marsh edges along the Atlantic Coastal
Plain from North Carolina south to Florida. It shares its
preference for this unique wetland habitat with the
similar but much larger Rare Skipper with which it
occasionally flies. Generally uncommon, it may occasionally be locally numerous when encountered. The
larvae pupate within a dense cocoon of silk and
leaves. Later instar larvae overwinter.

# Byssus Skipper
*Problema byssus*

**Family/Subfamily:** Skippers (Hesperiidae)/
Banded Skippers (Hesperiinae)

**Wingspan:** 1.4–1.8" (3.6–4.6 cm)

**Above:** bright orange with dark brown black borders,
dark veins and a black bar at the end of the forewing
cell; forewings pointed

**Below:** golden orange to yellow brown with a band of
paler yellow spots

**Sexes:** similar, although female has broader dark brown
borders and reduced golden orange scaling above

**Egg:** white, laid singly on the upper surface of host
leaves

**Larva:** bluish green with short white hairs and a light reddish brown head marked with pale yellow vertical
stripes

**Larval Host Plants:** various grasses including Big
Bluestem and Eastern Gammagrass

**Habitat:** forested wetlands and coastal marshes and
their margins

**Broods:** two generations

**Abundance:** rare to occasional

**Compare:** Rare Skipper (pg. 329) lacks faint ventral spot
band.

Resident

Jan. Feb. Mar. Apr. May June July Aug. Sept. Oct. Nov. Dec.

**Dorsal (above)**
thin black bar at
end of cell
dark borders

**Ventral (below)**
golden orange to
yellow brown
band of paler spots

**Comments:** A butterfly of contrasting color, it is a rich
  tawny orange below compared to a much darker sooty
  black dorsal appearance. Duke's Skipper is a denizen
  of shaded southern swamps and marshes, but may
  occasionally be encountered along wet roadside
  ditches if sedges are present. Typically rare and localized, it can be a difficult species to locate. Larvae
  overwinter. Much remains to be learned about its life
  history, ecology and behavior.

# Duke's Skipper
*Euphyes dukesi*

**Family/Subfamily:** Skippers (Hesperiidae)/
Banded Skippers (Hesperiinae)

**Wingspan:** 1.50–1.75" (3.8–4.4 cm)

**Above:** dark sooty brownish black with faint tawny
orange scaling along base of costal forewing margin;
wings are somewhat rounded

**Below:** rich brownish orange with black forewing base
and a pale ray through the hindwing

**Sexes:** similar, although female has small white forewing
spots

**Egg:** laid singly on the underside of host leaves

**Larva:** pale green; reddish brown head with a black spot
on the forehead surrounded by white

**Larval Host Plants:** various sedges, including shoreline
sedge

**Habitat:** shaded swamps and marshes and moist roadside ditches

**Broods:** two generations

**Abundance:** rare to occasional

**Compare:** Dion Skipper (pg. 311) has bright orange scaling above with more pointed forewings and prefers
more open, sunlit habitats.

Resident

Jan. Feb. Mar. Apr. May June July Aug. Sept. Oct. Nov. Dec.

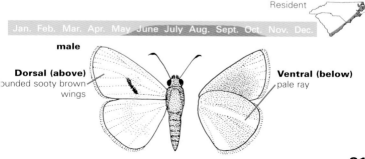

male

**Dorsal (above)**
rounded sooty brown
wings

**Ventral (below)**
pale ray

Male

Female | Winter | Larva

**Comments:** Much debate centers on the origin of the butterfly's common name. One interpretation points to its narrow black forewing spot, which to many observers resembles a closed eye. Others suggest that behavior is responsible. Late-season adults overwinter in reproductive diapause. They are sedentary and seldom seen, but often become active on mild days to nectar before disappearing again. The Sleepy Orange is a frequent garden visitor, and is most abundant in late summer and fall. It typically rests and feeds with its wings closed; orange color is seen primarily in flight or when the butterfly is basking in the sun.

# Sleepy Orange
*Eurema nicippe*

**Family/Subfamily:** Whites and Sulphurs (Pieridae)/ Sulphurs (Coliadinae)

**Wingspan:** 1.3–2.0" (3.3–5.1 cm)

**Above:** bright orange with broad irregular black wing borders; forewing cell bears small, elongated black spot

**Below:** hindwings seasonally variable; butter-yellow with brown markings in summer-form and tan to reddish brown with darker pattern elements in winter-form

**Sexes:** similar, although female is larger and less vibrant with heavier ventral hindwing pattern

**Egg:** white, laid singly on host leaves

**Larva:** green with thin, cream lateral stripe and numerous short hairs

**Larval Host Plants:** various wild sennas including Sicklepod Senna, Wild Senna, Maryland Senna and Coffee Senna

**Habitat:** open, disturbed sites including roadsides, vacant fields, agricultural land, parks and gardens

**Broods:** multiple generations

**Abundance:** common

**Compare:** Orange Sulphur (pg. 335) has rounded black cell spot on dorsal forewing, regular wing borders and distinct, red-rimmed silver spot on ventral hindwing.

Resident Visitor

Jan. Feb. Mar. Apr. May June July Aug. Sept. Oct. Nov. Dec.

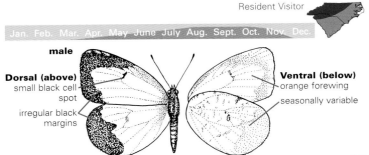

**male**

**Dorsal (above)**
small black cell spot

irregular black margins

**Ventral (below)**
orange forewing

seasonally variable

**321**

Male

Ventral

Larva

**Comments:** This large checkerspot is a spotty but often locally common resident, primarily limited to the western mountainous portions of the Carolinas. It is most frequently encountered in moist woodland openings or along forested streams in close proximity to its larval hosts. The Silvery Checkerspot tends to fluctuate considerably in abundance from year to year, being locally common at times or nearly absent. Adults are often seen nectaring at flowers or while perched on low vegetation. They have a slow, gliding flight and maneuver close to the ground in open areas or along woodland roads. Males often gather at damp soil or animal dung.

# Silvery Checkerspot
*Chlosyne nycteis*

**Family/Subfamily:** Brush-foots (Nymphalidae)/
True Brush-foots (Nymphalinae)

**Wingspan:** 1.4–2.0" (3.6–5.1 cm)

**Above:** tawny orange with black bands and wide black
borders; hindwing has a submarginal row of white-centered, somewhat square black spots

**Below:** hindwing is pale yellow brown with a dark marginal patch and incomplete marginal row of silvery
white crescents

**Sexes:** similar

**Egg:** cream, laid in large clusters on underside of leaves

**Larva:** dark brownish black; wide yellow orange lateral
band and several rows of black spines; black head

**Larval Host Plants:** various composites including
Wing-stem, Gravelweed, White Crownbeard, asters,
sunflowers, Cut-leaf Coneflower and Sneezeweed

**Habitat:** moist woodland openings, forest margins,
stream corridors, semi-open areas near moist deciduous forest, and mesic fields

**Broods:** multiple generations

**Abundance:** rare to locally common

**Compare:** Gorgone Checkerspot (pg. 291) has a distinctly zigzag pattern on ventral hindwing.

Resident

Jan. Feb. Mar. Apr. May June July Aug. Sept. Oct. Nov. Dec.

**Dorsal (above)**
submarginal row of
white-centered
black spots

**Ventral (below)**
incomplete row of
silvery white
marginal crescents

323

Male

Male

Larva

**Comments:** As its name suggests, the Northern Crescent is primarily a northern species. It ranges southward through the Appalachian Mountains into nearby Virginia and West Virginia. North Carolina specimens represent the presence of small, isolated colonies or rare strays. Extremely similar to the Pearl Crescent, many individuals may not be reliably differentiated in the field.

# Northern Crescent
*Phyciodes selenis*

**Family/Subfamily:** Brush-foots (Nymphalidae)/
True Brush-foots (Nymphalinae)

**Wingspan:** 1.5–1.9" (3.8–4.8 cm)

**Above:** tawny orange with fine black bands, spots and
broad wing borders

**Below:** hindwing is golden yellow brown with fine brown
markings and a dark marginal patch typically enclosing
a pale crescent (occasionally absent)

**Sexes:** similar

**Egg:** green, laid in clusters on the underside of host
leaves

**Larva:** dark pinkish brown with fine white mottling, a
dark dorsal line, a lateral cream stripe and numerous
short, pinkish gray branched spines

**Larval Host Plants:** various asters

**Habitat:** moist semi-open woodlands, stream margins

**Broods:** single generation

**Abundance:** rare

**Compare:** Pearl Crescent (pg. 301) is smaller, has less
extensive open orange areas on dorsal hindwing; often
lacks pale crescent on ventral hindwing.

Stray

Jan. Feb. Mar. Apr. May June July Aug. Sept. Oct. Nov. Dec.

**Dorsal (above)**
fine black markings

orange on
hindwing fairly
open

**Ventral (below)**
dark marginal patch;
often without pale
crescent

**325**

Male

Larva

**Comments:** This is a large, somewhat distinctive skipper often common in freshwater and brackish marshes along coastal portions of the Carolinas. Living up to its name, the species has noticeably rounded, somewhat wide wings. Although often locally abundant, much remains to be learned about the life history and ecology of this attractive but poorly known species. Adults are readily encountered at the blossoms of numerous wetland flowers.

# Broad-winged Skipper
*Poanes viator*

**Family/Subfamily:** Skippers (Hesperiidae)/
Banded Skippers (Hesperiinae)

**Wingspan:** 1.3–2.1" (3.3–5.3 cm)

**Above:** tawny orange with broad dark brown borders;
forewing has rounded apex; hindwing has dark brown
scaling along veins; male lacks stigma

**Below:** hindwing is tawny orange with distinct broad
pale ray through pale spot band

**Sexes:** similar, although female has cream white spots
on the forewing above

**Egg:** grayish, laid singly on the underside of host leaves

**Larva:** pale brown with an overall velvety appearance;
reddish brown head

**Larval Host Plants:** Giant Cutgrass, Annual Wild Rice
and Common Reed, also sedges

**Habitat:** forested swamps and associated clearings and
margins

**Broods:** two generations

**Abundance:** occasional to common

**Compare:** Yehl Skipper (pg. 309) lacks distinct broad
pale ray through spot band on hindwing below.

Resident

Jan. Feb. Mar. Apr. May June July Aug. Sept. Oct. Nov. Dec.

**male**

**Dorsal (above)**
squarish yellow
orange spots

lacks stigma

broad dark borders

**Ventral (below)**
broad pale ray

pale spot band

Larva

**Comments:** The Rare Skipper is a large golden orange butterfly entirely restricted to brackish coastal marshes. It occurs in small, highly localized and spotty colonies, but may be fairly common when encountered. A great many of its typical haunts are somewhat inaccessible to humans. Adults are best found and observed feeding at the numerous wildflowers growing along the marsh edges. Little detailed information is known about the life history, ecology and behavior of this skipper.

# Rare Skipper
*Problema bulenta*

**Family/Subfamily:** Skippers (Hesperiidae)/
Banded Skippers (Hesperiinae)

**Wingspan:** 1.5–2.1" (3.8–5.3 cm)

**Above:** bright orange with broad, dark brown borders
and a black bar at the end of the forewing cell;
forewings pointed

**Below:** unmarked golden orange; forewing has distinct
black cell-end bar often visible above hindwing

**Sexes:** similar, although female has brown wing bases
and reduced golden orange scaling

**Egg:** currently undocumented

**Larva:** pale green with a dark dorsal stripe; reddish
brown head with cream markings

**Larval Host Plants:** Giant Cutgrass and Wild Rice

**Habitat:** brackish coastal marshes

**Broods:** two generations

**Abundance:** rare to occasional

**Compare:** Byssus Skipper (pg. 317) is smaller and has a
faint spot band on the wings below.

Resident

Jan. Feb. Mar. Apr. May June July Aug. Sept. Oct. Nov. Dec.

**Dorsal (above)**
black cell-end bar
dark borders

**Ventral (below)**
black cell-end bar
unmarked golden
orange

**329**

**Comments:** The Palatka Skipper, also called the Sawgrass Skipper, is generally found in close proximity to its larval host. The large adults have a strong, rapid flight and frequently visit available flowers. Males perch on low vegetation and readily investigate passing objects. The larvae construct individual shelters on the host by tying leaves together with silk.

# Palatka Skipper
*Euphyes pilatka*

**Family/Subfamily:** Skippers (Hesperiidae)/ Banded Skippers (Hesperiinae)

**Wingspan:** 1.7–2.1" (4.3–5.3 cm)

**Above:** male is tawny orange with dark brown borders and narrow black forewing stigma; female is browner with reduced orange markings

**Below:** rust; forewing dark brown toward base and along trailing edge

**Sexes:** dissimilar; female is primarily brown with an orange spot band on the forewing

**Egg:** light green, laid singly on host leaves

**Larva:** green with fine black dots; head is whitish with three black stripes

**Larval Host Plants:** Sawgrass

**Habitat:** wetlands, coastal freshwater and brackish marshes

**Broods:** multiple generations

**Abundance:** occasional

**Compare:** Dion Skipper (pg. 311) has a pale ray through the center of the ventral hindwing.

Resident

Jan. Feb. Mar. Apr. May June July Aug. Sept. Oct. Nov. Dec.

male

**Dorsal (above)**
broad dark border

**Ventral (below)**
dull reddish brown

**Ventral**

**Larva**

**Comments:** The medium-sized Variegated Fritillary shares its affinity for open, sunny habitats with the similar-looking Gulf Fritillary but is generally less common and highly localized in occurrence. Adults have low, erratic flight but regularly stop to nectar at available flowers. It is an occasional garden visitor. A butterfly of the southern U.S., it regularly expands its range northward each summer.

# Variegated Fritillary
*Euptoieta claudia*

**Family/Subfamily:** Brush-foots (Nymphalidae)/
Longwing Butterflies (Heliconiinae)

**Wingspan:** 1.75–2.25" (4.4–5.7 cm)

**Above:** pale brownish orange with dark markings, narrow light median band and darker reddish orange base

**Below:** overall brown; forewing has basal orange scaling; hindwing mottled with tan, cream and dark brown; hindwing lacks silvery spots

**Sexes:** similar, although female is larger and has broader, more rounded wings

**Egg:** tiny cream eggs laid singly on host leaves and tendrils

**Larva:** reddish orange with black-spotted white stripes and black spines

**Larval Host Plants:** various violets and Passion Flower vines

**Habitat:** open, sunny sites including roadsides, pastures, old fields and utility easements

**Broods:** multiple generations

**Abundance:** occasional to locally abundant

**Compare:** Gulf Fritillary (pg. 351) is brighter orange with more elongated wings and silvery ventral spots.

Resident

Jan. Feb. Mar. Apr. May June July Aug. Sept. Oct. Nov. Dec.

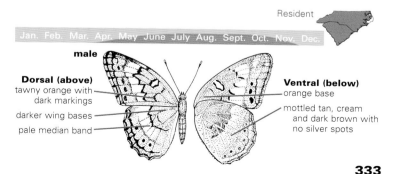

**male**

**Dorsal (above)**
tawny orange with dark markings

darker wing bases

pale median band

**Ventral (below)**
orange base

mottled tan, cream and dark brown with no silver spots

Larva

**Comments:** The Orange Sulphur can be found in virtually any open landscape and is particularly abundant in
and around cultivated alfalfa fields, where it occasionally can become a serious pest. Because of this strong
preference, it is often called the Alfalfa Butterfly. It is a
prolific colonizer. Adults have a rapid, erratic flight.
Butterflies produced very early or late in the season
are generally smaller and darker. This butterfly typically
rests and feeds with its wings closed, so its orange
color is seen primarily in flight or when the butterfly is
basking in the sun.

# Orange Sulphur
*Colias eurytheme*

**Family/Subfamily:** Whites and Sulphurs (Pieridae)/ Sulphurs (Coliadinae)

**Wingspan:** 1.6–2.4" (4.1–6.1 cm)

**Above:** bright yellow orange with black wing borders and black forewing cell spot; hindwing has central orange spot

**Below:** yellow with row of dark submarginal spots; hindwing has one or two central red-rimmed silvery spots

**Sexes:** similar, although female has yellow spots in broader black wing borders and are less vibrant; female is occasionally white

**Egg:** white, laid singly on host leaves

**Larva:** green with thin, cream lateral stripe and numerous short hairs

**Larval Host Plants:** Alfalfa, White Sweet Clover, White Clover and vetches

**Habitat:** open, sunny sites including roadsides, old fields, alfalfa fields, pastures and home gardens

**Broods:** multiple generations

**Abundance:** occasional to abundant

**Compare:** Southern Dogface (pg. 393) is larger, has pointed forewings and wide, scalloped forewing borders. Clouded Sulphur (pg. 395) lacks dorsal orange scaling.

Resident

Jan. Feb. Mar. Apr. May June July Aug. Sept. Oct. Nov. Dec.

**Dorsal (above)**
black cell spot
smooth black borders
bright yellow orange
central orange spot

**Ventral (below)**
red-rimmed silvery spots
dark submarginal spots

Ventral

Larva

**Comments:** The Painted Lady, a resident of northern Mexico, annually colonizes much of North America each summer. In the Carolinas, its occurrence is sporadic with populations that vary from year to year; it is seldom very common. The Painted Lady has a rapid, erratic flight, but frequently stops to perch or nectar. The larvae construct individual shelters of loose webbing on host leaves.

# Painted Lady
*Vanessa cardui*

**Family/Subfamily:** Brush-foots (Nymphalidae)/
True Brush-foots (Nymphalinae)

**Wingspan:** 1.75–2.40" (4.4–6.1 cm)

**Above:** pinkish orange with dark markings and small
white spots near tip of forewing

**Below:** brown with cream patches in ornate cobweb pattern; hindwing has row of four small eyespots and
marginal band of lavender spots

**Sexes:** similar

**Egg:** small pale green eggs laid singly on host leaves

**Larva:** variable; greenish yellow with black mottling to
charcoal with cream mottling and several rows of light-
colored, branched spines

**Larval Host Plants:** large variety of plants including
thistles and mallows, also Hollyhock

**Habitat:** open, disturbed sites including roadsides, old
fields, fallow agricultural land, pastures, utility easements and gardens

**Broods:** multiple generations

**Abundance:** occasional

**Compare:** American Painted Lady (pg. 339) is overall
more orange, has a more extended forewing apex and
two large ventral hindwing eyespots.

Visitor

Jan. Feb. Mar. Apr. May June July Aug. Sept. Oct. Nov. Dec.

**male**

**Dorsal (above)**
no small white spot
(as in American
Painted Lady)

connected black
markings

**Ventral (below)**
ornate cobweb patter

four submarginal
eyespots

Ventral

Larva

**Comments:** Considered a common butterfly, the American Painted Lady is often overlooked despite its attractiveness. The intricately detailed, agate-like design on the underside of the wings is a sharp contrast to the bold orange and black pattern above. Nervous and wary, it is difficult to approach and a challenge to closely observe. When disturbed, it takes off in a low, erratic flight but often returns to a nearby location just a few moments later. The larvae construct individual shelters on the host by spinning together leaves and flowerheads with silk. Inside, the larvae safely rest when not actively feeding.

# American Painted Lady
*Vanessa virginiensis*

**Family/Subfamily:** Brush-foots (Nymphalidae)/
True Brush-foots (Nymphalinae)

**Wingspan:** 1.75–2.40" (4.4–6.1 cm)

**Above:** orange with dark marks and borders; forewing has
small white spots near extended and squared-off apex

**Below:** brown with ornate, cream cobweb pattern; hindwing has two large eyespots and narrow lavender
marginal band

**Sexes:** similar; females with broader wings

**Egg:** small pale green eggs laid singly on upper surface
of host leaves

**Larva:** variable; greenish yellow with narrow black bands
to black with cream bands and numerous red-based,
branched spines; pair of prominent white spots on
each segment

**Larval Host Plants:** Purple Cudweed, Wandering
Cudweed, Narrow-leaved Cudweed, Sweet
Everlasting, pussy-toes and others

**Habitat:** open, disturbed sites including roadsides, old
fields, pastures, utility easements and gardens

**Broods:** multiple generations

**Abundance:** common to abundant

**Compare:** Painted Lady (pg. 337) is pinker and has a row
of four small ventral hindwing eyespots.

Resident

Jan. Feb. Mar. Apr. May June July Aug. Sept. Oct. Nov. Dec.

**Dorsal (above)**
apex extended and
squared off

small white spot

disconnected black
markings

some blue in one or
more eyespots

**Ventral (below)**
small white spot

ornate cobweb pattern

two large eyespots

**339**

Ventral

**Comments:** This uncommon, somewhat reclusive
species is named for the prominent green spots along
the margins of the heavily mottled underwings. The
Green Comma overwinters as an adult, hibernating in
log piles, tree hollows or even within manmade structures. As a result, the adults are long-lived and may
survive for many months. Adults have an erratic, darting flight and frequently perch on the ground or on
fallen logs or rocks. They are often quite wary and difficult to closely approach. It seldom visits flowers,
feeding instead at rotting fruit, dung, carrion and tree
sap.

# Green Comma
*Polygonia faunus*

**Family/Subfamily:** Brush-foots (Nymphalidae)/
True Brush-foots (Nymphalinae)

**Wingspan:** 1.9–2.4" (4.8–6.1 cm)

**Above:** tawny orange with irregular wing edge, heavy
black borders and spots; hindwing has a row of small
yellow orange spots in border and a short, stubby tail;
forewing apex is extended and squared off

**Below:** mottled brown with dark striations, small green
spots along margin and silver comma in center of hindwing; darker toward base; dead leaf appearance

**Sexes:** similar

**Egg:** green, laid singly on host leaves

**Larva:** variable; yellow brown to rust with transverse
black and yellow bands and spots, two wavy orange
lateral lines and white and orange branched spines

**Larval Host Plants:** gooseberry, willows and birch

**Habitat:** higher elevation deciduous and mixed forests,
riparian woodlands, woodland roads, forest edges and
adjacent open brushy areas

**Broods:** singe generation

**Abundance:** rare to occasional

**Compare:** Eastern Comma (pg. 343) has less jagged
wing margins and lacks submarginal green spots on
wings below.

Resident

Jan. Feb. Mar. Apr. May June July Aug. Sept. Oct. Nov. Dec.

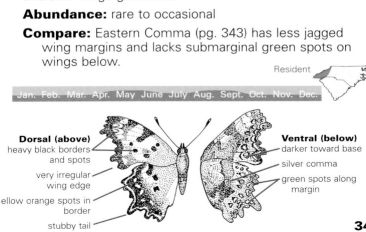

**Dorsal (above)**
heavy black borders
and spots

very irregular
wing edge

ellow orange spots in
border

stubby tail

**Ventral (below)**
darker toward base

silver comma

green spots along
margin

Summer

Ventral

Winter

Larva

**Comments:** The Eastern Comma survives the winter as an adult, hibernating in log piles, tree hollows or even within manmade structures. Once warm weather returns, they become active and reproduce, with the resulting generation flying in early summer. A woodland butterfly, it inhabits riparian forests, clearings and adjacent open, brushy areas. Adults have a rapid, erratic flight and frequently perch on overhanging branches or tree trunks. They are quite wary and difficult to closely approach. It seldom visits flowers, feeding instead at rotting fruit, dung, carrion and tree sap.

# Eastern Comma
*Polygonia comma*

**Family/Subfamily:** Brush-foots (Nymphalidae)/
True Brush-foots (Nymphalinae)

**Wingspan:** 2.0–2.4" (5.1–6.1 cm)

**Above:** tawny orange with black borders and spots, irregular wing edge; forewing apex is extended and squared off; hindwing has single distinct tail; summer-form hindwing is primarily black; winter-form has increased orange

**Below:** seasonally variable; dead leaf appearance; summer-form is heavily mottled light and dark brown; winter-form is uniform brown with fine striations; hindwing has a single distinct curved silvery spot in center

**Sexes:** similar

**Egg:** green, laid singly or in small stacks on host leaves

**Larva:** variable; black to greenish brown with a white lateral band and several rows of white, branched spines

**Larval Host Plants:** Canadian Wood Nettle, American Elm, False Nettle, Common Hop and nettles

**Habitat:** deciduous forests, riparian woodlands, forest roads and edges and adjacent open brushy areas

**Broods:** two generations

**Abundance:** occasional

**Compare:** Question Mark (pg. 349) is larger and has silver spots in the center of ventral hindwing shaped like a question mark.

Resident

Jan. Feb. Mar. Apr. May June July Aug. Sept. Oct. Nov. Dec.

**Dorsal (above)**
black spots

seasonally variable;
orange with black
spots to primarily
black

tail

**Ventral (below)**
seasonally variable;
dead leaf
appearance

silver comma

Ventral

Larva

**Comments:** This woodland butterfly is aptly named for its dull gray, highly striated underwings that closely resemble tree bark. Infrequent and somewhat reclusive, the Gray Common occurs in small, highly localized colonies. It is most often encountered as a single, isolated individual along woodland trails, clearings or roadsides. Adults have an erratic, darting flight and frequently perch tree trunks or on overhanging branches. They are quite wary and difficult to closely approach. Both sexes seldom visit flowers, but prefer to feed at rotting fruit, dung, carrion and tree sap.

# Gray Comma
*Polygonia progne*

**Family/Subfamily:** Brush-foots (Nymphalidae)/
True Brush-foots (Nymphalinae)

**Wingspan:** 2.1–2.4" (5.3–6.1 cm)

**Above:** tawny orange with heavy black borders and
spots, highly irregular wing edge; forewing apex is
extended and squared off; hindwing bears a single
short, stubby tail; hindwing border encloses a row of
small yellow orange spots

**Below:** dull gray with fine striations and a narrow silver
comma in center of hindwing; darker toward base

**Sexes:** similar

**Egg:** green, laid singly on host leaves

**Larva:** variable; yellow brown with transverse black lines
between segments, dark botches and dashes, and
several rows of pale yellow and black branched spines

**Larval Host Plants:** Appalachian Gooseberry, Wild
Azalea and perhaps Paper Birch

**Habitat:** deciduous woodlands, clearings, woodland
roads, forest edges and adjacent open brushy areas

**Broods:** two generations

**Abundance:** rare to occasional

**Compare:** Green Comma (pg. 341) has distinct row of
green spots on ventral hindwing margin.

Resident

Jan. Feb. Mar. Apr. May June July Aug. Sept. Oct. Nov. Dec.

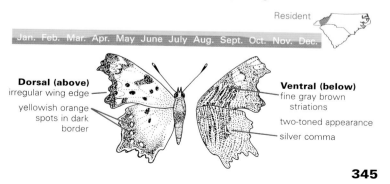

**Dorsal (above)**
irregular wing edge

yellowish orange
spots in dark
border

**Ventral (below)**
fine gray brown
striations

two-toned appearance

silver comma

**345**

Male

Ventral | Winter | Larva

**Comments:** Named after one of its larval hosts, the Goatweed Butterfly has a leaf-like pattern on the wings below. It is a rare stray or occasional temporary colonist in the Carolinas and is seldom encountered in large numbers. Adults have a strong, rapid flight and perch with their wings closed on the trunks of trees, branches or on the ground and can be quite a challenge to find, let alone closely approach. The butterfly prefers rotting fruit and sap over flower nectar. It produces distinct seasonal forms. This butterfly typically rests and feeds with its wings closed, so its orange color is seen primarily in flight or when the butterfly is basking in the sun.

# Goatweed Butterfly
*Anaea andria*

**Family/Subfamily:** Brush-foots (Nymphalidae)/
Leafwings (Charaxinae)

**Wingspan:** 2.25–3.00" (5.7–7.6 cm)

**Above:** pointed forewings and hindwing tail; male is
bright reddish orange; female is lighter orange with
dark markings and pale band along wing margins

**Below:** seasonally variable; brownish gray resembling a
dead leaf; winter-form is more heavily patterned, has
longer hindwing tails and a more pronounced hooked
forewing apex

**Sexes:** similar, although female is lighter with more
extensive dorsal markings

**Egg:** gray-green, laid singly on host leaves

**Larva:** gray-green with light head and numerous tiny light
spots

**Larval Host Plants:** Silver Croton

**Habitat:** dry pinelands, woodland edges and adjacent
open areas

**Broods:** two generations

**Abundance:** rare

**Compare:** Question Mark (pg. 349) has irregular wing
edges, an extended and squared-off forewing apex
and black dorsal forewing spots.

Stray

Jan. Feb. Mar. Apr. May June July Aug. Sept. Oct. Nov. Dec.

male

**Dorsal (above)**
hooked apex

tail

**Ventral (below)**
dead leaf appearance

**Summer**

**Ventral** **Winter** **Larva**

**Comments:** The Question Mark's dark markings, irregular wing edges and cryptically colored undersides help the butterfly resemble a dead leaf when resting. Its name comes from the silvery marks on the ventral hindwing that resemble a rudimentary question mark. Adults have a strong, rapid flight but frequently alight on overhanging branches, tree trunks or leaf litter. Wary and nervous, they are often difficult to closely approach. Males readily establish territories and aggressively investigate any passing objects. Both sexes visit rotting fruit, dung, carrion and tree sap.

# Question Mark
*Polygonia interrogationis*

**Family/Subfamily:** Brush-foots (Nymphalidae)/ True Brush-foots (Nymphalinae)

**Wingspan:** 2.25–3.00" (5.7–7.6 cm)

**Above:** orange with black spots, narrow lavender borders and irregular, jagged edges; seasonally variable; summer-form hindwing is primarily black; winter-form has increased orange on hindwing

**Below:** seasonally variable; pinkish brown dead leaf appearance; hindwing has two, small median silvery spots that form a question mark

**Sexes:** similar

**Egg:** green, laid singly or in small groups on top of each other on host leaves

**Larva:** gray to black with orange and cream stripes and spots and several rows of branched spines

**Larval Host Plants:** Sugarberry, American Elm, Winged Elm, also False Nettle and Stinging Nettle

**Habitat:** deciduous forests, moist woodlands, forest edges and adjacent open areas

**Broods:** two generations

**Abundance:** occasional

**Compare:** Eastern Comma (pg. 343) is smaller and has comma-shaped spot on ventral hindwing.

Resident

Jan. Feb. Mar. Apr. May June July Aug. Sept. Oct. Nov. Dec.

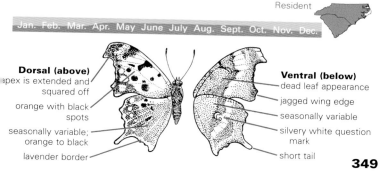

**Dorsal (above)**
apex is extended and squared off

orange with black spots

seasonally variable; orange to black

lavender border

**Ventral (below)**
dead leaf appearance

jagged wing edge

seasonally variable

silvery white question mark

short tail

Female

Ventral

Larva

**Comments:** The Gulf Fritillary's hindwings below are beautifully decorated with numerous silvery, mercury-like patches that quickly distinguish it from the equally common Monarch. In mild years, it might survive the winter along extreme southern coastal portions of South Carolina, but is typically a seasonal colonist. It stops to nectar at colorful flowers and is an abundant garden visitor. Adults have a low, rapid flight. It is one of several migratory species in the Southeast. During the fall, adults migrate southward in large numbers and overwinter in southern portions of Florida.

# Gulf Fritillary
*Agraulis vanillae*

**Family/Subfamily:** Brush-foots (Nymphalidae)/
Longwing Butterflies (Heliconiinae)

**Wingspan:** 2.5–3.0" (6.4–7.6 cm)

**Above:** bright orange elongated wings with black markings and three small, black-rimmed white spots in
forewing cell; hindwing has chain-like black border

**Below:** brown with elongated silvery spots

**Sexes:** similar, although female is paler above with
increased black markings

**Egg:** yellow eggs laid singly on host leaves

**Larva:** orange with greenish black stripes and black,
branched spines

**Larval Host Plants:** various Passion Flower vines
including Maypop, Corky-Stemmed Passion Flower and
Yellow Passion Flower and Incense Passion Flower

**Habitat:** open, disturbed sites including roadsides, old
fields, utility easements, parks and gardens

**Broods:** multiple generations

**Abundance:** occasional to common

**Compare:** The Gulf Fritillary's distinct ventral silver spots
quickly distinguish it from the Variegated Fritillary (pg.
333) and Monarch (pg. 363).

Resident Visitor

Jan. Feb. Mar. Apr. May June July Aug. Sept. Oct. Nov. Dec.

**male**

**Dorsal (above)**
elongated wings

black-rimmed white
spots

chain-like black border

**Ventral (below)**
orange forewing base

elongated silvery spots

Male

Ventral

Larva

**Comments:** The colorful Viceroy is usually found in wetland areas close to its larval host. It has a quick, gliding flight and is often wary and difficult to closely approach. Males perch on overhanging branches and occasionally dart out to explore their territory or investigate passing objects. Once thought to be a palatable mimic of the Monarch, studies now indicate that both species are toxic to certain predators.

# Viceroy
*Limenitis archippus*

**Family/Subfamily:** Brush-foots (Nymphalidae)/ Admirals and Relatives (Limenitidinae)

**Wingspan:** 2.6–3.2" (6.6–8.1 cm)

**Above:** orange with black markings, veins and broad wing borders; borders contain central row of small, white spots; forewing has black postmedian band and white spots; hindwing has distinct thin, black post-median line

**Below:** as above with lighter orange coloration and increased white markings

**Sexes:** similar

**Egg:** gray-green, laid singly on tip of host leaves

**Larva:** mottled green, brown and cream with two long, knobby horns on thorax

**Larval Host Plants:** various willows including Black Willow, Carolina Willow, Weeping Willow

**Habitat:** pond edges, wetlands, roadside ditches and moist areas supporting willows

**Broods:** multiple generations

**Abundance:** occasional to common

**Compare:** Monarch (pg. 363) is larger and lacks black postmedian hindwing line. Gulf Fritillary (pg. 351) is brighter orange with elongated wings and distinct silver ventral spots.

Resident

Jan. Feb. Mar. Apr. May June July Aug. Sept. Oct. Nov. Dec.

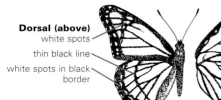

**Dorsal (above)**
white spots
thin black line
white spots in black border

**Ventral (below)**
orange with black bands containing white spots

Ventral

**Comments:** This widespread northern species is restricted to the extreme western portions of the Carolinas. The species is much less common and more localized than its abundant relative, the Great Spangled Fritillary, with which it is often confused. A butterfly of open areas, it frequents old fields, meadows, open woodlands and forested roadsides where it is an avid flower visitor. Young larvae overwinter and complete development the following spring.

# Aphrodite Fritillary
*Speyeria aphrodite*

**Family/Subfamily:** Brush-foots (Nymphalidae)/ Longwing Butterflies (Heliconiinae)

**Wingspan:** 2.7–3.3" (6.9–8.4 cm)

**Above:** bright orange with heavy black lines and spots

**Below:** variable; hindwing is dark orange brown to reddish brown with numerous large metallic silver spots and a narrow yellowish submarginal band

**Sexes:** similar, although female is larger with darker brown wing bases, more pronounced black markings and paler golden orange wings

**Egg:** tiny cream eggs laid singly and somewhat haphazardly near host leaves

**Larva:** velvety black with several rows of black spines; lateral two rows of spines reddish brown with black tips

**Larval Host Plants:** various violets

**Habitat:** open woodlands, fields, moist meadows, pastures and roadsides

**Broods:** single generation

**Abundance:** uncommon to occasional

**Compare:** Great Spangled Fritillary (pg. 359) is slightly larger and has wide yellowish submarginal band on the hindwing below.

Resident

Jan. Feb. Mar. Apr. May June July Aug. Sept. Oct. Nov. Dec.

**Dorsal (above)**
bright orange with heavy black lines and spots

**Ventral (below)**
large metallic silver spots

reddish brown base

narrow yellowish submarginal band

**Female**

**Ventral**

**Larva**

**Comments:** The Queen, a year-round resident of the Deep South, regularly expands its range north to colonize coastal portions of the Carolinas. Adults have a slow, soaring flight and are fond of flowers. It is a frequent garden visitor and typically rests and feeds with wings closed. The larvae feed on plants in the Milkweed family and sequester various chemicals that render them and the resulting adults highly distasteful to certain predators. The rich mahogany and black coloration advertises their unpalatability. Unlike the Monarch, it doesn't undertake annual migrations.

# Queen
*Danaus gilippus*

**Family/Subfamily:** Brush-foots (Nymphalidae)/ Milkweed Butterflies (Danainae)

**Wingspan:** 3.0–3.5" (7.6–8.9 cm)

**Above:** mahogany with black wing borders and small white forewing spots; male has a single black post-median androconial scent patch on the hindwing

**Below:** reddish brown; forewing as above; hindwing has black veins and small white spots in black border

**Sexes:** similar; female lacks black scent patch

**Egg:** white, laid singly on host leaves

**Larva:** black with transverse cream stripes and yellow spots; body has three pairs of long, black filaments

**Larval Host Plants:** various milkweed family plants including Mexican Milkweed, White Swamp Milkweed, Sandhill Milkweed and Swamp Milkweed

**Habitat:** savannas, pastures, pinelands, roadsides, old fields, utility easements, coastal areas and gardens

**Broods:** multiple generations

**Abundance:** occasional

**Compare:** Monarch (pg. 363) is larger, brighter orange and has black wing veins above. Gulf Fritillary (pg. 351) has elongated wings and prominent silver ventral spots.

Visitor

Jan. Feb. Mar. Apr. May June July Aug. Sept. Oct. Nov. Dec.

**male**

**Dorsal (above)**
mahogany in color
white spots
veins not outlined in black
scent patch on males only

**Ventral (below)**
black veins on hindwing
double row of white spots

**357**

Female

Female    Larva

**Comments:** This is the most abundant and widespread true fritillary in the Carolinas. Adults have a strong, directed flight but readily stop to nectar at the blossoms of colorful wildflowers, being particularly fond of milkweed, Joe-pye weed and thistles. It is not uncommon to see several individuals at a single flower. The butterfly has a single, protracted summer flight with males typically emerging several weeks before females. Although females may be on the wing in early July, they do not begin laying eggs until later and often fly until late September. Young larvae overwinter and complete development the following spring.

# Great Spangled Fritillary
*Speyeria cybele*

**Family/Subfamily:** Brush-foots (Nymphalidae)/ Longwing Butterflies (Heliconiinae)

**Wingspan:** 2.9–3.8" (7.1–9.7 cm)

**Above:** bright orange with heavy black lines and spots

**Below:** variable; hindwing is dark orange brown to brown with numerous large metallic silver spots and a broad yellowish submarginal band

**Sexes:** similar, although female is larger with darker brown wing bases, more pronounced black markings and paler golden orange to pale yellowish wings

**Egg:** tiny cream eggs laid singly and somewhat haphazardly on or host leaves

**Larva:** velvety black with several rows of reddish orange-based black spines

**Larval Host Plants:** various violets

**Habitat:** open, sunny sites including roadsides, pastures, old fields, wet meadows and utility easements close to wooded sites

**Broods:** single generation

**Abundance:** occasional to locally abundant

**Compare:** Aphrodite Fritillary (pg. 355) is smaller, less common and has a narrow yellowish submarginal band on the ventral hindwing.

Resident

Jan. Feb. Mar. Apr. May June July Aug. Sept. Oct. Nov. Dec.

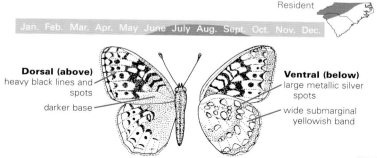

**Dorsal (above)**
heavy black lines and spots
darker base

**Ventral (below)**
large metallic silver spots
wide submarginal yellowish band

Female

Ventral

Larva

**Comments:** This beautiful butterfly was once found within North Carolina. Sadly though, it has been extirpated from the state. It is possible, but highly unlikely, that a few small remnant colonies still remain. The Regal Fritillary is of strong conservation concern as it is rapidly declining over most of its once widespread range. Young larvae overwinter and complete development the following spring.

# Regal Fritillary
*Speyeria idalia*

**Family/Subfamily:** Brush-foots (Nymphalidae)/ Longwing Butterflies (Heliconiinae)

**Wingspan:** 3.1–4.0" (7.9–10.2 cm)

**Above:** forewing is bright reddish orange with black markings; hindwing is black with a bluish cast; male has an outer row of orange spots and a inner row of white spots on hindwing

**Below:** hindwing is dark brown with numerous black-edged white (not silvery) spots

**Sexes:** similar, although female is larger with both rows of hindwing spots white

**Egg:** tiny cream eggs laid singly and somewhat haphazardly near host leaves

**Larva:** gray black; ochre yellow to reddish orange dorsal band and similar colored lateral mottling; dorsal spines are silvery white with black tips while those along the sides have orange bases

**Larval Host Plants:** Bird's-foot Violet

**Habitat:** wet fields, marshes, meadows and grasslands

**Broods:** single generation

**Abundance:** extirpated; possible colonies still remain

**Compare:** unique

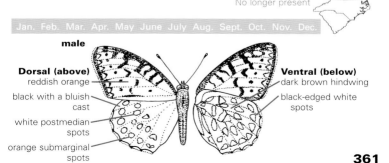

No longer present

Jan. Feb. Mar. Apr. May June July Aug. Sept. Oct. Nov. Dec.

male

Dorsal (above)
reddish orange

black with a bluish cast

white postmedian spots

orange submarginal spots

Ventral (below)
dark brown hindwing

black-edged white spots

**361**

**Female**

**Ventral** **Larva**

**Comments:** The Monarch is undoubtedly the most familiar and widely recognized butterfly in North America. Its annual fall mass migration is one of the greatest natural events undertaken by any organism on Earth. Adults have a strong, soaring flight and are abundant garden visitors. The striped larvae feed on plants in the Milkweed family from which they sequester toxic chemicals that render them and the resulting adults distasteful to certain predators. The adult butterflies advertise this unpalatability in dramatic fashion with their bold orange and black coloration.

# Monarch
*Danaus plexippus*

**Family/Subfamily:** Brush-foots (Nymphalidae)/
Milkweed Butterflies (Danainae)

**Wingspan:** 3.5–4.0" (8.9–10.2 cm)

**Above:** orange with black veins and wing borders; black
borders have two rows of small white spots; male has
small black androconial scent patch on center of hindwing

**Below:** as above with lighter orange coloration

**Sexes:** similar; female lacks black scent patch

**Egg:** white, laid singly on host leaves

**Larva:** white with transverse black and yellow stripes;
there is a pair of long, black filaments on each end

**Larval Host Plants:** various milkweeds including
Mexican Milkweed, White Swamp Milkweed, Sandhill
Milkweed, Swamp Milkweed and Common Milkweed

**Habitat:** open, sunny locations including old fields, roadsides, pinelands, utility easements, sandhills, fallow
agricultural land and gardens

**Broods:** multiple generations

**Abundance:** occasional to common

**Compare:** Queen (pg. 357) is smaller, deep mahogany in
color and lacks black wing veins above. Gulf Fritillary
(pg. 351) has elongated wings and prominent silver
ventral spots.

Resident

Jan. Feb. Mar. Apr. May June July Aug. Sept. Oct. Nov. Dec.

**male**

**Dorsal (above)**
white spots
bright orange
two rows of white
spots
black veins
scent patch on
males only

**Ventral (below)**
lighter yellow orange

Male

Female pg. 71    Male    Larva

**Comments:** The Diana Fritillary is a showstopper! Strikingly sexually dimorphic, the blue and black female mimics the toxic Pipevine Swallowtail to presumably gain protection from predators. Although often spotty and localized, it can be fairly numerous when encountered. Adults have a strong, directed flight but readily stop to nectar at colorful wildflowers, being particularly fond of milkweeds, Joe-pye Weed and thistles. As with other fritillaries, males typically emerge several weeks before females. Young larvae overwinter and complete development the following spring.

# Diana Fritillary
*Speyeria diana*

**Family/Subfamily:** Brush-foots (Nymphalidae)/ Longwing Butterflies (Heliconiinae)

**Wingspan:** 3.5–4.4" (8.9–11.2 cm)

**Above:** male is dark unmarked blackish brown with bright orange on the outer third; female is black basally with white and iridescent blue spots on outer half

**Below:** forewing is orange with heavy black markings toward base; male hindwing is brownish orange with two rows of small narrow silver dashes; female hindwing is chocolate brown

**Sexes:** dissimilar; female is black basally with white and iridescent blue spots on outer half; ventral hindwing is chocolate brown

**Egg:** tiny cream eggs laid singly and somewhat haphazardly near host leaves

**Larva:** velvety black with several rows of reddish orange based black spines

**Larval Host Plants:** various woodland violets

**Habitat:** rich, moist deciduous mountain woodlands, stream corridors, forested roads, clearings and adjacent open areas

**Broods:** single generation

**Abundance:** occasional to locally abundant

**Compare:** unique

Resident

Jan. Feb. Mar. Apr. May June July Aug. Sept. Oct. Nov. Dec.

**male**

**Dorsal (above)**
black
orange
(female primarily black and blue)

**Ventral (below)**
black markings toward base

lacks prominent silvery hindwing spots characteristic of other fritillaries

**365**

Ventral    Larva

**Comments:** New data suggests that the once abundant
Common Checkered-Skipper has been almost entirely
displaced by the White Checkered-Skipper's recent
expansion into the Deep South. This trend is continuing northward along the coast into South Carolina.
Virtually identical in appearance and behavior, the two
sibling species can be reliably differentiated only
through dissection. Adults have a rapid, darting flight.
While at rest, they generally hold their wings open,
allowing a glimpse of the butterfly's distinctive
checkerboard pattern.

# Common/White Checkered-Skipper
*Pyrgus communis/Pyrgus albescens*

**Family/Subfamily:** Skippers (Hesperiidae)/
Spread-wing Skippers (Pyrginae)

**Wingspan:** 0.75–1.25" (1.9–3.2 cm)

**Above:** male is black with numerous small, white spots
and some bluish white scaling on base of wings and
thorax; female is dark brown with reduced white scaling

**Below:** white with tan to brown irregular bands and
spots

**Sexes:** similar, although female has reduced white
markings

**Egg:** pale green, laid singly on host leaves

**Larva:** gray-green with dark dorsal stripe, light side
stripes and black head

**Larval Host Plants:** various mallow family plants
including Indian Hemp, Broomweed, Poppy Mallow
and False Mallow

**Habitat:** open, disturbed sites including roadsides, old
fields, utility easements and fallow agricultural land

**Broods:** multiple generations

**Abundance:** common

**Compare:** Tropical Checkered-Skipper (pg. 369) male has
increased bluish white scaling on wings above and
both sexes have more diffuse or smudged ventral
bands.

Resident

Jan. Feb. Mar. Apr. May June July Aug. Sept. Oct. Nov. Dec.

**Dorsal (above)**
fringe checkered to
apex

generally black and
white above

marginal spots much
smaller than those
in submarginal row

**Ventral (below)**
distinct bands

paler below

**367**

Male

Female | Ventral | Larva

**Comments:** The Tropical Checkered-Skipper, while common in Florida, is a rare stray in the Carolinas. Adults have a fast, erratic flight and buzz rapidly among low vegetation, pausing occasionally to nectar or perch with their wings outstretched. The light green larvae construct individual shelters on the host by tying two or more leaves together with silk.

# Tropical Checkered-Skipper
*Pyrgus oileus*

**Family/Subfamily:** Skippers (Hesperiidae)/
Spread-wing Skippers (Pyrginae)

**Wingspan:** 0.90–1.35" (2.3–3.4 cm)

**Above:** male is dark blackish brown with numerous
small, white spots and extensive bluish gray hair-like
scales on base of wings and thorax; female is dark
brown with reduced white scaling

**Below:** powdery white with tan to brown irregular,
somewhat diffuse or smudged bands

**Sexes:** similar, although female darker with reduced
white markings

**Egg:** pale green, laid singly on host leaves

**Larva:** yellow-green with thin green dorsal stripe, light
side stripes and black head; body is covered with
numerous small, light-colored hairs

**Larval Host Plants:** various mallows including Indian
Hemp, Broomweed, Poppy Mallow and False Mallow

**Habitat:** open, disturbed sites including roadsides, old
fields, utility easements and fallow agricultural land

**Broods:** multiple generations

**Abundance:** rare

**Compare:** Common/White Checkered-Skipper (pg. 367)
male has reduced bluish white dorsal scaling; both
sexes have well-defined ventral bands.

Stray

Jan. Feb. Mar. Apr. May June July Aug. Sept. Oct. Nov. Dec.

**male**

**Dorsal (above)**
fringe dark toward
apex

long blue gray hair-
like scales

marginal spots
similar in size to
submarginal
spots

**Ventral (below)**
dark spots along
leading margin

diffuse, smudged
bands

Summer

Winter pg. 391

Male

Larva

**Comments:** The Barred Sulphur is a common butterfly of open, weedy sites throughout the Deep South. In the Carolinas, it is a seasonal colonist. Adults have a quick, dancing flight and typically bob among low vegetation. Like many other sulphurs, the butterfly displays dramatic seasonal variation in color and behavior. Individuals produced during the summer months are short-lived, reproductively active and almost immaculate white beneath. Winter-forms have highly pigmented and patterned ventral hindwings and over-winter in reproductive diapause. It typically rests and feeds with its wings closed.

# Barred Sulphur
*Eurema daira*

**Family/Subfamily:** Whites and Sulphurs (Pieridae)/ Sulphurs (Coliadinae)

**Wingspan:** 1.0–1.8" (2.5–4.6 cm)

**Above:** seasonally variable; winter-form bright yellow; summer-form whitish; all have black wing tips and a black bar along the trailing edge of forewing; forewing bar is diffuse or absent in female; wide black hindwing border is reduced to a small patch in winter-form

**Below:** seasonally variable; hindwings light gray to pure white in summer-form; tan to reddish brown with numerous pattern elements in winter-form

**Sexes:** similar, although black markings paler and less extensive in female

**Egg:** white, laid singly on host leaves

**Larva:** green with thin, lateral cream stripe

**Larval Host Plants:** Joint Vetch, possibly other Fabaceae

**Habitat:** open, disturbed sites including roadsides, pastures, utility easements, vacant fields, pond margins

**Broods:** multiple generations

**Abundance:** uncommon to occasional

**Compare:** Dainty Sulphur (pg. 387) is much smaller and lacks white ventral hindwings. Little Sulphur (pg. 389) lacks black dorsal forewing bar.

Visitor

Jan. Feb. Mar. Apr. May June July Aug. Sept. Oct. Nov. Dec.

**male**

**Dorsal (above)**
black apex

large black bar

irregular black marginal patch or continuous border

**Ventral (below)**
seasonally variable; immaculate white to patterned reddish brown

**371**

**Comments:** The West Virginia White is strictly a woodland species. Flying only in early spring, and often
highly localized, the butterfly rarely ventures far from
populations of its larval host. It is a poor pioneer
species, being reluctant to cross large expanses of
open habitat to colonize nearby scattered patches of
available woodland or second growth forest. Adults
have a low, weak flight and maneuver slowly along the
forest floor, stopping occasionally to nectar on a variety of early season wildflowers. Males frequently
puddle at damp ground along woodland trails or roads.

# West Virginia White
*Pieris virginiensis*

**Family/Subfamily:** Whites and Sulphurs (Pieridae)/ Whites (Pierinae)

**Wingspan:** 1.2–1.6" (3.0–4.1 cm)

**Above:** unmarked white to smoky white with black scaling along costal margin and wing bases

**Below:** hindwings white with brown gray scaling along veins

**Sexes:** similar, although female typically appears more smoky gray on the wings above

**Egg:** greenish white, laid singly on underside of leaves

**Larva:** gray green with longitudinal yellow orange stripes and covered with small black dots

**Larval Host Plants:** toothwort, rockcress, Garlic Mustard and bittercress

**Habitat:** rich, moist deciduous woodlands, forest edges, and adjacent meadows or open areas

**Broods:** single generation

**Abundance:** occasional to common; often quite localized in occurrence

**Compare:** The Cabbage White (pg. 379) is similar in size but has charcoal wing tips, one or two postmedian black spots on the dorsal forewing and lacks brown gray scaling along the veins on the ventral hindwing.

Resident

Jan. Feb. Mar. Apr. May June July Aug. Sept. Oct. Nov. Dec.

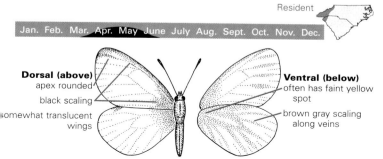

**Dorsal (above)**
apex rounded

black scaling

somewhat translucent wings

**Ventral (below)**
often has faint yellow spot

brown gray scaling along veins

Male

Ventral   Female   Larva

**Comments:** This small, delicate butterfly provides a welcome hint of approaching spring by appearing during the cool, unpredictable early season. Aptly named, adult Falcate Orangetip may be quickly identified by their characteristic hooked forewings and orange wingtips in males. Although considered a forest species, it avoids dense shade and is typically associated with open, thin woods and adjacent clearings. Adults have a quick flight and maneuver adeptly close to the ground. Both sexes visit a variety of early spring flowers and readily nectar on their larval hosts.

# Falcate Orangetip
*Anthocharis midea*

**Family/Subfamily:** Whites and Sulphurs (Pieridae)/ Whites (Pierinae)

**Wingspan:** 1.25–1.75" (3.2–4.4 cm)

**Above:** white; forewing has a small black cell spot and distinctive hooked (falcate) tip; male has orange tips on forewings; female has white forewing tips

**Below:** hindwings white; heavily marbled with gray

**Sexes:** dissimilar; male has orange forewing tips

**Egg:** orange, laid singly on leaves, stems or flower buds

**Larva:** blue-green with a narrow yellow orange dorsal stripe, bright white lateral stripe and small black dots

**Larval Host Plants:** herbaceous plants in mustard family, including rockcress, Field Peppergrass, Cut-leaf Toothwort, bittercress, Shepherd's Purse and Garlic Mustard

**Habitat:** lowland forested stream corridors, hilltops, roadsides, open deciduous forests and pine-oak woodlands, and associated margins and clearings

**Broods:** single generation

**Abundance:** uncommon to common

**Compare:** West Virginia White (pg. 373) lacks black forewing cell spot, falcate wing tips and heavily marbled ventral hindwing.

Resident

Jan. Feb. Mar. Apr. May June July Aug. Sept. Oct. Nov. Dec.

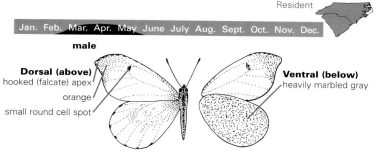

male

**Dorsal (above)**
hooked (falcate) apex

orange

small round cell spot

**Ventral (below)**
heavily marbled gray

**375**

Male

Female    Larva

**Comments:** A butterfly of dry, weedy sites, the Checkered White is a wide-ranging species but often only locally abundant. Adults have a fast, erratic flight and can be difficult to approach. The sexes differ dramatically and can be easily told apart even from a distance. Adults produced in early spring or late fall, under cooler conditions and shorter days, are typically smaller, darker and more heavily patterned. It readily stops to nectar on low flowers and is especially fond of the blossoms of its larval hosts.

# Checkered White
*Pontia protodice*

**Family/Subfamily:** Whites and Sulphurs (Pieridae)/ Whites (Pierinae)

**Wingspan:** 1.25–2.00" (3.2–5.1 cm)

**Above:** male is white with charcoal markings on forewing and immaculate hindwing; female is grayish white with extensive black or grayish brown checkered markings on both wings

**Below:** hindwings white with grayish markings and yellow-green scaling along the veins; seasonally variable; cool season individuals are more heavily patterned

**Sexes:** similar, although female has more extensive black markings

**Egg:** yellow, laid singly on host leaves or flowers

**Larva:** gray; longitudinal yellow-orange stripe, black dots

**Larval Host Plants:** Virginia Peppergrass, Common Pepperweed, Garden Yellowrocket, Shepherd's Purse

**Habitat:** open, disturbed sites including roadsides, pastures, utility easements, vacant fields, agricultural land

**Broods:** multiple generations

**Abundance:** occasional to common

**Compare:** Cabbage White (pg. 379) has charcoal wing tips and single or double dorsal forewing spots. Great Southern White (pg. 381) has scalloped black forewing border.

Resident Visitor

| Jan. | Feb. | Mar. | Apr. | May | June | July | Aug. | Sept. | Oct. | Nov. | Dec. |

**male**

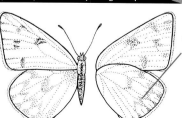

**Dorsal (above)**
white with black checkered pattern

**Ventral (below)**
seasonally variable; cool season forms are more heavily patterned

**377**

Larva

**Comments:** Accidentally introduced into the U.S. from Europe in 1860, the Cabbage White (or European Cabbage Butterfly) quickly radiated across much of North America. It is one of the few butterfly species that is considered to be a serious agricultural and garden pest. Adults have a slow, somewhat awkward flight and are easy to observe. Although appearing pure white from a distance, upon closer inspection the wings beneath are delicately shaded with yellow.

# Cabbage White
*Pieris rapae*

**Family/Subfamily:** Whites and Sulphurs (Pieridae)/ Whites (Pierinae)

**Wingspan:** 1.5–2.0" (3.8–5.1 cm)

**Above:** male is white with single black postmedian forewing spot and wing tips; female is white with two black postmedian forewing spots

**Below:** forewing white with two black spots and yellow tips; hindwing immaculate whitish yellow

**Sexes:** similar, although female has two black spots on forewing

**Egg:** white, laid singly on host leaves and flowers

**Larva:** green with small lateral yellow dashes and numerous short hairs

**Larval Host Plants:** cultivated and wild members of the Mustard Family including Virginia Peppergrass, Wild Mustard, Wild Radish, broccoli, cabbage, turnip, cauliflower and radish

**Habitat:** open, disturbed sites including roadsides, old fields, utility easement, agricultural land and gardens

**Broods:** multiple generations

**Abundance:** occasional to common

**Compare:** Checkered White (pg. 377) lacks black tip on forewing. Great Southern White (pg. 381) is larger and has scalloped black forewing borders.

Resident

Jan. Feb. Mar. Apr. May June July Aug. Sept. Oct. Nov. Dec.

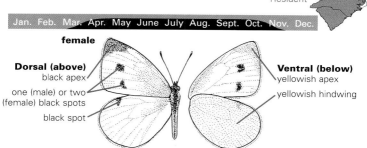

**female**

**Dorsal (above)**
black apex
one (male) or two (female) black spots
black spot

**Ventral (below)**
yellowish apex
yellowish hindwing

**Male**

**Female** **Larva**

**Comments:** The Great Southern White's antennae have a distinctive blue tip. While males are always white, females may be white, gray or an intermediate color. Adults have a low, generally casual flight, but can move swiftly if disturbed. The butterfly occasionally experiences tremendous population outbreaks. Extensive northward movements of adults along the coast often follow such explosions in number.

# Great Southern White
*Ascia monuste*

**Family/Subfamily:** Whites and Sulphurs (Pieridae)/
Whites (Pierinae)

**Wingspan:** 1.5–2.5" (3.8–6.4 cm)

**Above:** male is immaculate white with narrow, black
scalloped forewing border; female is seasonally variable; dirty white to gray with more pronounced black
forewing border and black forewing cell spot

**Below:** males white with pale yellow hindwings; females
dirty white to gray

**Sexes:** similar, although female has more gray scaling
and a single black dorsal forewing spot

**Egg:** yellow, laid singly or in small groups on host leaves

**Larva:** yellow with grayish purple longitudinal stripes,
black dots and numerous small hairs

**Larval Host Plants:** Virginia Peppergrass, Saltwort, Sea
Rocket and others

**Habitat:** open sites including roadsides, coastal dunes,
salt marshes, vacant fields, agricultural land

**Broods:** multiple generations

**Abundance:** rare to uncommon

**Compare:** Cabbage White (pg. 379) is smaller and lacks
black scalloped forewing border.

Stray

Jan. Feb. Mar. Apr. May June July Aug. Sept. Oct. Nov. Dec.

**male**

**Dorsal (above)**
narrow, scalloped
black border

blue tips

**Ventral (below)**

**381**

Ventral

**Comments:** The White Peacock has a fast, erratic flight usually within a few feet of the ground and readily alights on low vegetation. They avidly visit available flowers, where they can be easily observed. A year-round resident of southern Florida, it annually disperses north and occasionally makes its way to portions of South Carolina. Its occurrence is spotty and varies from year to year.

# White Peacock
*Anartia jatrophae*

**Family/Subfamily:** Brush-foots (Nymphalidae)/
True Brush-foots (Nymphalinae)

**Wingspan:** 2.0–2.5" (5.1–6.4 cm)

**Above:** white with dark brown marking and dull orange
scaling along wing margins; forewing has a single
black spot; hindwing has a short, stubby tail and two
black postmedian spots

**Below:** as above with paler brown wing margins and
pinkish markings

**Sexes:** similar, though female is larger and has broader
wings

**Egg:** green, laid singly on host leaves

**Larva:** charcoal gray with small silvery-white spots and
black, branched spines

**Larval Host Plants:** Frogfruit and Water Hyssop

**Habitat:** open, disturbed sites including wet ditches,
roadsides, pond edges, old fields and gardens

**Broods:** multiple generations

**Abundance:** rare to occasional

**Compare:** unique

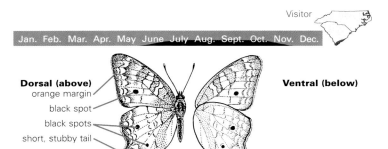

Visitor

| Jan. | Feb. | Mar. | Apr. | May | June | July | Aug. | Sept. | Oct. | Nov. | Dec. |

**Dorsal (above)**
orange margin
black spot
black spots
short, stubby tail

**Ventral (below)**

White-form female

Male pg. 395

Larva

**Comments:** Like its close relative the Orange Sulphur, this aggressive colonizer has likely benefitted from the spread of agriculture. A butterfly of open, sunny habitats, the Clouded Sulphur can be commonly encountered. It is generally most abundant in commercial clover or alfalfa fields, where it may occasionally become a serious pest. It is extremely variable in both size and color. Both yellow and white form females are common and adults produced under cool, early spring or late fall conditions tend to be smaller and somewhat darker.

# Clouded Sulphur
*Colias philodice*

**Family/Subfamily:** Whites and Sulphurs (Pieridae)/ Sulphurs (Coliadinae)

**Wingspan:** 1.90–2.75" (4.8–7.0 cm)

**Above:** clear lemon yellow with bold, solid black wing borders and a prominent black forewing cell spot; hindwing has a central orange spot

**Below:** yellow to greenish yellow; row (often faint or occasionally absent) of dark submarginal spots; pink wing fringes; hindwing has central red-rimmed silvery spot and adjacent smaller satellite spot

**Sexes:** dissimilar; female is less vibrant, often has more black scaling and broader black borders enclosing yellow spots; also has a common white form

**Egg:** white, laid singly on host leaves

**Larva:** blue-green with a lateral cream stripe marked in black below and often containing faint red dashes

**Larval Host Plants:** various weedy herbaceous legumes including Alfalfa, Red Clover, White Clover and White Sweet Clover

**Habitat:** open, sunny sites including roadsides, fields, alfalfa fields, parks, vacant lots, pastures, gardens

**Broods:** multiple generations

**Abundance:** occasional to abundant

**Compare:** White-form female may not be reliably distinguished in the field.

Resident

Jan. Feb. Mar. Apr. May June July Aug. Sept. Oct. Nov. Dec.

male

**Dorsal (above)**
black spot
yellow with no orange scaling
smooth black border
orange spot

**Ventral (below)**
two silvery spots
pink fringe
black dusting on hindwing
partial postmedian row of dark spots

**385**

Male

Larva

**Comments:** As its name implies, the Dainty Sulphur is our smallest sulphur. It flies low among the vegetation and is easily overlooked. Although found throughout much of Florida, the Dainty Sulphur is a rare vagrant or seasonal colonist northward into the Carolinas. It produces distinct seasonal forms that vary considerably in ventral hindwing coloration. Cooler season forms are more heavily pigmented.

## Dainty Sulphur
*Nathalis iole*

**Family/Subfamily:** Whites and Sulphurs (Pieridae)/ Sulphurs (Coliadinae)

**Wingspan:** 0.75–1.25" (1.9–3.2 cm)

**Above:** lemon yellow with black forewing tip and black bar along trailing edge of forewing; female has orange-yellow hindwings with more extensive black markings

**Below:** hindwings yellow with greenish markings; seasonally variable; winter-form more heavily pigmented

**Sexes:** similar, although black markings more extensive on female

**Egg:** yellow, laid singly on host leaves

**Larva:** green with thin, lateral yellow and lavender stripes

**Larval Host Plants:** Spanish Needles and Green Carpetweed

**Habitat:** open, disturbed sites including roadsides, pastures, utility easements, vacant fields, agricultural land and canal banks

**Broods:** multiple generations

**Abundance:** rare

**Compare:** Barred Sulphur (pg. 391) is larger with white ventral hindwings; lacks orange along costal margin of ventral forewing.

Stray

Jan. Feb. Mar. Apr. May June July Aug. Sept. Oct. Nov. Dec.

**male**

**Dorsal (above)**
black apex
diffuse black bar

**Ventral (below)**
black spots
orange scaling
seasonally variable; yellow olive to heavily dusted with black

Male

Female

Larva

**Comments:** The Little Sulphur has a low, darting flight
and an affinity for dry, open habitats. Although found
throughout the year in the Deep South, it regularly colonizes much of the East each year and is particularly
abundant in the late summer and early fall. Like other
members of the genus, it produces different seasonal
forms that vary in coloration, behavior and reproductive
activity.

# Little Sulphur
*Eurema lisa*

**Family/Subfamily:** Whites and Sulphurs (Pieridae)/ Sulphurs (Coliadinae)

**Wingspan:** 1.0–1.6" (2.5–4.1 cm)

**Above:** bright yellow with black forewing tip, narrow black wing borders and often a faint cell spot; female pale yellow to near white with lighter black markings

**Below:** seasonally variable; hindwings yellow to near white with pinkish red spot on outer margin (often absent on male) and several small, subtle spots or patches; winter-form darker yellow with additional pattern elements and pink wing fringe

**Sexes:** similar, although markings and color paler on female

**Egg:** white, laid singly on host leaves

**Larva:** green with thin, lateral cream-white stripe

**Larval Host Plants:** Partridge Pea and Sensitive Pea

**Habitat:** open, disturbed sites including roadsides, pastures, utility easements, vacant fields, agricultural land and open sandhills

**Abundance:** occasional to common

**Compare:** Dainty Sulphur (pg. 387) and Barred Sulphur (pg. 391) both have black dorsal forewing bar. Dainty has black postmedian spots on ventral forewing.

Resident

Jan. Feb. Mar. Apr. May June July Aug. Sept. Oct. Nov. Dec.

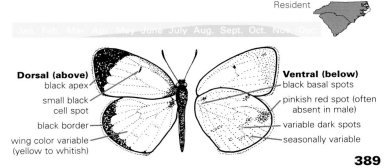

**Dorsal (above)**
- black apex
- small black cell spot
- black border
- wing color variable (yellow to whitish)

**Ventral (below)**
- black basal spots
- pinkish red spot (often absent in male)
- variable dark spots
- seasonally variable

**389**

Male

Summer pg. 371    Winter    Larva

**Comments:** The Barred Sulphur is a common butterfly of open, weedy sites throughout the Deep South. In the Carolinas, it is a seasonal colonist. Adults have a quick, dancing flight and typically bob among low vegetation. Like many other sulphurs, the butterfly displays dramatic seasonal variation in color and behavior. Individuals produced during the summer months are short-lived, reproductively active and almost immaculate white beneath. Winter-forms have highly pigmented and patterned ventral hindwings and overwinter in reproductive diapause. It typically rests and feeds with its wings closed.

# Barred Sulphur
*Eurema daira*

**Family/Subfamily:** Whites and Sulphurs (Pieridae)/ Sulphurs (Coliadinae)

**Wingspan:** 1.0–1.8" (2.5–4.6 cm)

**Above:** seasonally variable; winter-form bright yellow; summer-form whitish; all have black wing tips and a black bar along the trailing edge of forewing; forewing bar is diffuse or absent in female; wide black hindwing border is reduced to a small patch in winter-form

**Below:** seasonally variable; hindwings light gray to pure white in summer-form; tan to reddish brown with numerous pattern elements in winter-form

**Sexes:** similar, although black markings paler and less extensive in female

**Egg:** white, laid singly on host leaves

**Larva:** green with thin, lateral cream stripe

**Larval Host Plants:** Joint Vetch, possibly other Fabaceae

**Habitat:** open, disturbed sites including roadsides, pastures, utility easements, vacant fields, pond margins

**Broods:** multiple generations

**Abundance:** uncommon to occasional

**Compare:** Dainty Sulphur (pg. 387) is much smaller and lacks white ventral hindwings. Little Sulphur (pg. 389) lacks black dorsal forewing bar.

Visitor

Jan. Feb. Mar. Apr. May June July Aug. Sept. Oct. Nov. Dec.

**male**

**Dorsal (above)**
black apex
large black bar
irregular black marginal patch or continuous border

**Ventral (below)**
seasonally variable; immaculate white to patterned reddish brown

**391**

**Winter**  **Larva**

**Comments:** The only Carolina sulphur with pointed forewings, the Dogface is a medium-sized butterfly with a powerful, rapid flight. It is named after the unique pattern formed by the black and yellow markings on the wings above that resemble (with some imagination) the head of a dog in profile. Adults display considerable seasonal variation. Winter-forms are particularly stunning with rich pink scaling on the wings beneath. They overwinter in reproductive diapause.

# Southern Dogface
*Zerene cesonia*

**Family/Subfamily:** Whites and Sulphurs (Pieridae)/ Sulphurs (Coliadinae)

**Wingspan:** 1.9–2.5" (4.8–6.4 cm)

**Above:** yellow; forewing pointed with broad, black margin highly scalloped to form image of dog's head in profile, single back cell spot and increased black scaling on basal area; hindwing has narrow black margin

**Below:** hindwings seasonally variable; summer-form is yellow with two small silver spots; winter-form has increased rosy-pink scaling

**Sexes:** similar, although black markings duller and less extensive on female

**Egg:** white, laid singly on host leaves

**Larva:** variable; plain green to green with orange lateral stripe and transverse black and yellow stripes

**Larval Host Plants:** False Indigo, Summer Farewell

**Habitat:** open, dry sites including pastures, pineland, oak scrub, old fields and roadsides

**Broods:** multiple generations

**Abundance:** rare to occasional

**Compare:** All other Carolinas sulphurs lack the pointed forewing apex.

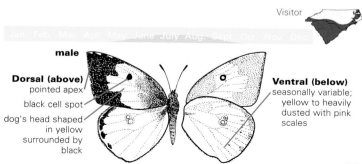

Visitor

Jan. Feb. Mar. Apr. May June July Aug. Sept. Oct. Nov. Dec.

male

**Dorsal (above)**
pointed apex

black cell spot

dog's head shaped in yellow surrounded by black

**Ventral (below)**
seasonally variable; yellow to heavily dusted with pink scales

**393**

Male

White-form female
pg. 385

Larva

**Comments:** Like its close relative the Orange Sulphur, this aggressive colonizer has likely benefitted from the spread of agriculture. A butterfly of open, sunny habitats, the Clouded Sulphur can be commonly encountered. It is generally most abundant in commercial clover or alfalfa fields, where it may occasionally become a serious pest. It is extremely variable in both size and color. Both yellow and white form females are common and adults produced under cool, early spring or late fall conditions tend to be smaller and somewhat darker.

# Clouded Sulphur
*Colias philodice*

**Family/Subfamily:** Whites and Sulphurs (Pieridae)/ Sulphurs (Coliadinae)

**Wingspan:** 1.90–2.75" (4.8–7.0 cm)

**Above:** clear lemon yellow with bold, solid black wing borders and a prominent black forewing cell spot; hind-wing has a central orange spot

**Below:** yellow to greenish yellow; row (often faint or occasionally absent) of dark submarginal spots; pink wing fringes; hindwing has central red-rimmed silvery spot and adjacent smaller satellite spot

**Sexes:** dissimilar; female is less vibrant, often has more black scaling and broader black borders enclosing yellow spots; also has a common white form

**Egg:** white, laid singly on host leaves

**Larva:** blue-green with a lateral cream stripe marked in black below and often containing faint red dashes

**Larval Host Plants:** various weedy herbaceous legumes including Alfalfa, Red Clover, White Clover and White Sweet Clover

**Habitat:** open, sunny sites including roadsides, fields, alfalfa fields, parks, vacant lots, pastures, gardens

**Broods:** multiple generations

**Abundance:** occasional to abundant

**Compare:** Orange Sulphur (pg. 335) always has at least some orange scaling above.

Resident

Jan. Feb. Mar. Apr. May June July Aug. Sept. Oct. Nov. Dec.

male

**Dorsal (above)**
black spot

yellow with no orange scaling

smooth black border

orange spot

**Ventral (below)**
two silvery spots

pink fringe

black dusting on hindwing

partial postmedian row of dark spots

Male

Female

Larva

**Comments:** The Cloudless Sulphur is a large yellow butterfly with a fast, powerful flight. Abundant throughout
the Southeast, it colonizes much of the Carolinas each
year. In favorable years, it survives the winter along
coastal parts of southern South Carolina. Adults have
an extremely long proboscis and can feed at many
long, tubular flowers inaccessible to other butterflies.
Fall individuals undergo a massive, southward migration
into Florida to overwinter. The annual event is one of
the Southeast's most impressive natural phenomena.

# Cloudless Sulphur
*Phoebis sennae*

**Family/Subfamily:** Whites and Sulphurs (Pieridae)/ Sulphurs (Coliadinae)

**Wingspan:** 2.2–2.8" (5.6–7.1 cm)

**Above:** unmarked bright lemon yellow; female has broken black wing borders and black forewing spot

**Below:** male is greenish yellow with virtually no markings; female is yellow with pinkish brown markings and several small silver spots in center of each wing; seasonally variable; winter-form adults more heavily marked

**Sexes:** similar, although female is more heavily marked

**Egg:** white, laid singly on host leaves or flower buds

**Larva:** green or yellow with broad lateral yellow stripe marked with blue spots or transverse bands

**Larval Host Plants:** various Cassia species including Partridge Pea, Sensitive Pea, Sicklepod Senna, Wild Senna, Maryland Senna and Christmas Senna

**Habitat:** open, disturbed sites including roadsides, vacant fields, agricultural land, parks and gardens

**Broods:** multiple generations

**Abundance:** occasional to abundant

**Compare:** Clouded Sulphur (pg. 395) and Orange Sulphur (pg. 335) have a solid black cell spot on the forewing, and broad, solid black wing margins.

Resident Visitor

Jan. Feb. Mar. Apr. May June July Aug. Sept. Oct. Nov. Dec.

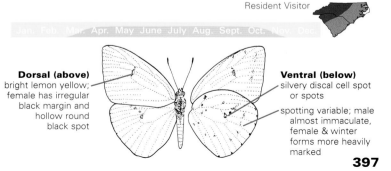

**Dorsal (above)**
bright lemon yellow; female has irregular black margin and hollow round black spot

**Ventral (below)**
silvery discal cell spot or spots

spotting variable; male almost immaculate, female & winter forms more heavily marked

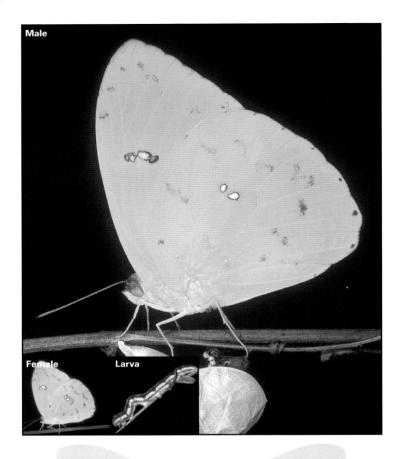

Male

Female Larva

**Comments:** Named for its distinct orange forewing band in males, the Orange-barred Sulphur is a large yellow butterfly with a strong and rapid flight. Found year-round in south Florida, it regularly moves northward each year to temporarily colonize additional areas. It is a rare stray in the Carolinas. Unlike many butterfly species, it is at home in more urban locations such as parks, gardens and even landscape nurseries where its ornamental larval hosts are commonly found. Adults are fond of brightly colored, long tubular blossoms.

# Orange-barred Sulphur
*Phoebis philea*

**Family/Subfamily:** Whites and Sulphurs (Pieridae)/ Sulphurs (Coliadinae)

**Wingspan:** 2.75–3.40" (7.0–8.6 cm)

**Above:** male is bright yellow with wide orange forewing bar and orange scaling along trailing edge of hindwing; female is seasonally variable with dark forewing borders and a broken diagonal spot band from the forewing apex

**Below:** male is bright yellow with small dark spots; summer-form female is orange yellow with fine dark mottling; winter-form female has increased markings

**Sexes:** dissimilar; female seasonally variable

**Egg:** white, laid singly on host leaves

**Larva:** green-yellow with a black lateral stripe and numerous black points

**Larval Host Plants:** various woody sennas including Candle Plant, Bahana Senna and Christmas Senna

**Habitat:** open, disturbed sites and urban areas

**Broods:** multiple generations

**Abundance:** rare

**Compare:** Cloudless Sulphur (pg. 397) is smaller and lighter yellow; males lack distinct dorsal orange bands and females lack the dark diagonal spots on forewing.

Stray

Jan. Feb. Mar. Apr. May June July Aug. Sept. Oct. Nov. Dec.

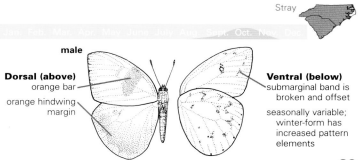

**male**

**Dorsal (above)**
orange bar

orange hindwing
margin

**Ventral (below)**
submarginal band is
broken and offset

seasonally variable;
winter-form has
increased pattern
elements

**399**

Male

Dark-form female
pg. 75

Female

Larva

**Comments:** Easily recognized by its bold, black stripes and yellow wings, the Tiger Swallowtail is one of the state's most familiar butterflies. Adults have a strong, agile flight and often soar high in the treetops. A common and conspicuous garden visitor, adults are readily drawn to available flowers. Males often congregate in large numbers at mud puddles or moist ground. Dark-form females mimic the toxic Pipevine Swallowtail to gain protection from predators.

# Eastern Tiger Swallowtail
*Papilio glaucus*

**Family/Subfamily:** Swallowtails (Papilionidae)/ Swallowtails (Papilioninae)

**Wingspan:** 3.5–5.5" (8.9–14.0 cm)

**Above:** yellow with black forewing stripes and broad black wing margins; single row of yellow spots along outer edge of each wing

**Below:** yellow with black stripes and black wing margins; hindwing margins have increased blue scaling and a single submarginal row of yellow-orange, crescent-shaped spots; abdomen yellow with black stripes

**Sexes:** dissimilar; male always yellow but females have two color forms; yellow female has increased blue scaling in black hindwing border; dark-form female is mostly black with extensive blue hindwing markings

**Egg:** green, laid singly on upper surface of host leaves

**Larva:** green; enlarged thorax and two small false eyespots

**Larval Host Plants:** Black Cherry, Wild Cherry, ash and Tulip Tree

**Habitat:** mixed forests, wooded swamps, forest edges, suburban gardens, old fields

**Broods:** multiple generations

**Abundance:** occasional to common

**Compare:** Yellow-form unique.

Resident

Jan. Feb. Mar. Apr. May June July Aug. Sept. Oct. Nov. Dec.

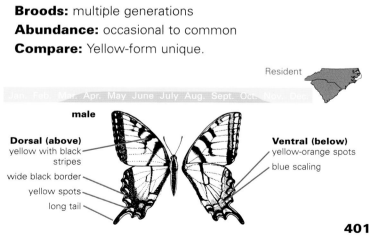

male

**Dorsal (above)**
yellow with black stripes

wide black border

yellow spots

long tail

**Ventral (below)**
yellow-orange spots

blue scaling

401

# BUTTERFLY SOCIETIES

Butterfly societies are a great way to learn more about butterflies and moths in the Carolinas. They also provide resources to make valuable contacts, share your enthusiasm and get out into the field.

Carolina Butterfly Society www.carolinabutterflysociety.org

Southern Lepidopterists' Society www.southernlepsoc.org

# PLANTS FOR YOUR BUTTERFLY GARDEN

The following are some recommended adult nectar sources and larval hosts for a Carolina butterfly garden. Most of the species are readily available at most retail garden centers or native plant nurseries. Before purchasing any landscape plant, always inquire with your local nursery personnel as to the specific soil, light, and care requirements needed for optimal growth and maintenance. Additionally, it is also a good idea to understand the plant's growth habit and eventual size at maturity before placing it in the ground. Visit a demonstration garden, a neighbor's yard or a nearby botanical garden to see how the plant looks after it has had a chance to grow a bit. Finally, remember that pesticides are not recommended for any butterfly garden as they can harm the very organisms you wish to attract. Consider using beneficial insects or insecticidal soap first before resorting to more extreme measures. If pesticides are required, always treat pest problems on a local level by applying treatment only to the infested plant and being careful of drift to neighboring vegetation.

## Adult Nectar Sources

### Herbaceous Perennials

Mexican Milkweed *(Asclepias curassavica)*

White Swamp Milkweed *(Asclepias perennis)*

Butterfly Weed *(Asclepias tuberosa)*

Carolina Aster *(Aster caroliniana)*

Spanish Needles *(Bidens alba)*

Florida Paint Brush *(Carphephorus corymbosus)*

Glorybower *(Clerodendron bungei)*

False Heather *(Cuphea hyssopifolia)*

Mist Flower *(Eupatorium coelestinum)*

Joe-pye Weed *(Eupatorium fistulosum)*

Indian Blanket *(Gaillardia pulchella)*

Moss Verbena *(Glandularia pulchella)*

Creeping Heliotrope *(Heliotropium amplexicaule)*

ironweed *(Vernonia spp.)*
Lantana *(Lantana camara)*
Purple Weeping Lantana *(Lantana montevidensis)*
blazing star *(Liatris spp.)*
Scarlet Beebalm *(Monarda didyma)*
Horse Mint *(Monarda punctata)*
Fire Spike *(Odontonema strictum)*
Drummond Phlox *(Phlox drummondii)*
Frogfruit *(Phylla nodiflora)*
Pickerelweed *(Pontederia cordata)*

mountain mint *(Pycnanthemum spp.)*
Autumn Sage *(Salvia gregii)*
Mexican Sage *(Salvia leucantha)*
sedum *(Sedum spp.)*
goldenrod *(Solidago spp.)*
Stokes Aster *(Stokesia laevis)*
Tall Verbena *(Verbena bonariensis)*

## Shrubs and Trees

Glossy Abelia *(Abelia x grandiflora)*
Red Buckeye *(Aesculus pavia)*
Silktree *(Albizia julibrissin)*
False Indigo *(Amorpha fruticosa)*
Butterfly Bush *(Buddleia davidii)*
Pride of Barbados *(Caesalpina pulcherima)*
Bluebeard *(Caryopteris x chandonensis)*
New Jersey Tea *(Ceanothus americanus)*
Buttonbush *(Cephalanthus occidentalis)*
Redbud *(Cercis canadensis)*
Sweet Pepperbush *(Clethra alnifolia)*

Firebush *(Hamelia patens)*
hibiscus *(Hibiscus spp.)*
Dahoon Holly *(Ilex cassine)*
Yaupon Holly *(Ilex vomitoria)*
Virginia Willow *(Itea virginica)*
Chickasaw Plum *(Prunus angustifolia)*
Black Cherry *(Prunus serotina)*
Hog Plum *(Prunus umbellata)*
Wild Azalea *(Rhododendron canescens)*
Blue Porterweed *(Stachytarpheta urticifolia)*
Sweet Viburnum *(Viburnum odoratissimum)*
Chaste Tree *(Vitex agnus-castus)*

## Vines

Coral Vine *(Antigon leptopus)*
Bougainvillea *(Bougainvillea glabra)*
morning glory *(Ipomoea spp.)*

Climbing Hempweed *(Mikania scandens)*
Maypop *(Passiflora incarnata)*
Flame Vine *(Senecio confuses)*

## Annuals

Ageratum *(Ageratum houstonianum)*
Spider Plant *(Cleome hasslerana)*
Dianthus *(Dianthus chinensis)*
Globe Amaranth *(Gomphrena globosa)*
impatiens *(Impatiens spp.)*
Sweet Alyssum *(Lobularia maritima)*

Flowering Tobacco *(Nicotiana alata)*
Pentas *(Pentas lanceolata)*
Drummond Phlox *(Phlox drummondii)*
Tropical Sage *(Salvia coccinea)*
Mexican sunflower *(Tithonia spp.)*
Verbena *(Verbena X hybrida)*
Zinnia *(Zinnia elegans)*

# Larval Host Plants

acanthus family (Acanthaceae)

Alfalfa (Medicago sativa)

Alligator Flag (Thalia geniculata)

Amaranth, Spiny (Amaranthus spinosus)

Arrowwood, Southern (Viburnum dentatum)

Ash, Prickly (Zanthoxylum americanum)

Ash, Wafer (Ptelea trifoliata)

ashes (Fraxinus spp.)

Aspen, Quaking (Populus tremuloides)

aspens (Populus spp.)

Aster, Bushy (Aster dumosus)

Aster, Frost (Aster pilosus)

Aster, Smooth Blue (Aster laevis)

Aster, Waxyleaf (Aster undulatus)

asters (Aster spp.)

Azalea, Flame (Rhododendron calendulaceum)

Azalea, Wild (Rhododendron nudilforum)

Barnyardgrass (Echinochloa crus-galli)

Bay, Red (Persea borbonia)

bean family (Fabacea)

beans (Phaseolus spp.)

Beech, American (Fagus gandiflora)

beggarweeds (Desmodium spp.)

bentgrass (Agrostis spp.)

Birch, Paper (Betula papyrifera)

birches (Betula spp.)

bittercress (Cardamine spp.)

Bitternut (Juglans cinerea)

Blackbead (Pithecellobium keyense)

blueberries (Vaccinium spp.)

Blueberry, Blue Ridge (Vaccinium vacillans)

Blueberry, Highbush (Vaccinium corymbosum)

bluegrass (Poa spp.)

Bluegrass, Kentucky (Poa pratensis)

Bluestem, Big (Andropogon gerardii)

Bluestem, Little (Schizachyrium scoparium)

bluestems (Andropogon spp.)

Bride's Feathers (Aruncus dioicus)

Bristlegrass, Hooked (Setaria verticillata)

Broomweed (Sida acuta)

Canarygrass, Reed (Phalaris arundinacea)

Candle Plant (Cassia alata)

Cane, Giant (Arundinaria gigantea)

Cane, Switch (Arundinaria gigantea spp. tecta)

cannas (Canna spp.)

Caper, Limber (Capparis flexuosa)

Carpetgrass, Broadleaf (Axonopus compressus)

Carpetweed, Green (Mollugo verticillata)

carrot family (Apiaceae) includes dill, fennel and parsley

Cassia species (Cassia spp.)

Cat's Claw (Pithecellobium unguis-cati)

Cedar, Atlantic White (Chamaecyparis thyoides)

Cedar, Southern Red (Juniperus silicicola)

Cherry, Black (Prunus serotina)

Cherry, Wild (Prunus virginiana)

Chestnut, American (Castanea dentata)

Cinquefoil, Canada (Potentilla canadensis)

Clover, Red (Trifolium pratense)

Clover, White (Trifolium repens)

Clover, White Sweet (Melilotus alba)

clovers (Trifolium spp.)

clovers, bush (Lespedeza spp.)

Cohosh, Black (Cimicifuga racemosa)

composites (Asteraceae others?XX)

Coneflower, Cut-leaf (Rudbecia

lanciniata)
Coontie (Zamia pumila)
Cordgrass, Smooth (Spartina alterniflora var. glabra)
cordgrass (Spartina spp.)
Crabgrass, Carolina (Digitaria cognata)
Crabgrass, Hairy (Digitaria sanguinalis)
Crabgrass, Shaggy (Digitaria villosa)
Crabgrass, Slender (Digitaria filiformis)
Croton, Silver (Croton argyranthemus)
Croton, Woolly (Croton linearis)
Crownbeard, White (Verbesina virginica)
Cudweed, Purple (Gnaphalium purpureum)
Cudweed, Narrow-leaved (Gnaphalium falcatum)
Cudweed, Wandering (Gnaphalium pensylvanicum)
Cutgrass, Giant (Zizaniopsis milacea)
Cutgrass, Rice (Leersia oryzoides)
Cutgrass, Southern (Leersia hexandra)
Deertongue (Dichanthelium clandestinum)
Dock, Curly (Rumex crispus)
Dogwood, Flowering (Cornus florida)
dogwoods (Cornus spp.)
Dutchman's Pipe (Aristolochia macrophylla)
Elm, American (Ulmus americana)
Elm, Winged (Ulmus alata)
elms (Ulmus spp.)
Fescue, Red (Festuca rubra)
Fig, Short-leaved (Ficus citrifolia)
Fig, Strangler (Ficus aurea)
Foldwing, Branched (Dicliptera brachiata)
foxglove, false (Agalinus spp.)
Frogfruit (Phyla nodiflora)
Gammagrass, Eastern (Tripsacum dactyloides)

glasswort (Salicornia spp.)
goatsbeard (Aruncus dioicus)
gooseberries (Ribes spp.)
Gooseberry, Appalachian (Ribes rotundifolium)
Grass, Bermuda (Cynodon dactylon)
Grass, Indian (Sorghastrum nutans)
Grass, Purpletop (Tridens flavus)
Grass, Salt (Distichlis spicata)
Grass, St. Augustine (Stenotaphrum secundatum)
Grass, Stiff-leaved Bear (Yucca filamentosa)
Grass, Switch (Panicum virgatum)
Grass, Timothy (Phleum pratense)
grasses (Poaceae)
Gravelweed (Verbesina helianthoides)
Ground Nut (Apios americana)
Hackberry, Common (Celtis occidentalis)
Hackberry, Dwarf (Celtis tenuifolia)
hackberries (Celtis spp.)
hawthorn (Crataegus spp.)
Hazelnut, Beaked (Corylus cornuta)
heath family (Ericaceae)
Hemp, Indian (Sida rhombifolia)
Hercules Club (Zanthoxylum clava-hercules)
Hickory, Bitternut (Carya cordiformis)
Hickory, Pignut (Carya glabra)
Hickory, Shagbark (Carya ovata)
hickories (Carya spp.)
Holly, American (Ilex opaca)
Holly, Dahoon (Ilex cassine)
Holly, Yaupon (Ilex vomitoria)
hollies (Ilex spp.)
Hollyhock (Althaea rosea)
Hop, Common (Humulus lupulus)
Horseflyweed (Baptisia tinctoria)
Huckleberry, Black (Gaylussacia baccata)
Hyssop, Water (Bacopa monnieri)
Indian Shot (Canna indica)
Indigo, Blue Wild (Baptisia australis)

Indigo, Carolina (*Indigofera caroliniana*)
Indigo, Creeping (*Indigofera spicata*)
Indigo, False (*Amorpha fruticosa*)
Indigo, Hairy (*Indigofera hirsuta*)
Indigo, White Wild (*Baptisia alba*)
Indigo, Wild (*Baptisia tinctoria*)
Kudzu (*Pueraria lobata*)
Lamb's Quarters (*Chenopidium album*)
Leadwort (*Plumbago auriculata*)
Leatherleaf (*Chamaedaphne calyculata*)
legumes (*Fabaceae*)
Locust, Black (*Robinia pseudoacacia*)
Lousewort, Canadian (*Pedicularis canadensis*)
lovegrass (*Eragrostis spp.*)
Lupine, Wild (*Lupinus perennis*)
lupin (*Lupinus spp.*)
mallow family (*Malvaceae spp.*)
Mallow, False (*Malvasreum corchorifolium*)
Mallow, Poppy (*Callirhoe papaver*)
Mango (*Mangifera indica*)
Mangrove, Black (*Avicennia germinans*)
Mangrove, Red (*Rhizophora mangle*)
mannagrass (*Glyceria spp.*)
Mannagrass, Fowl (*Glyceria striata*)
Maypop (*Passiflora incarnata –passionflower*)
milk peas (*Galactia spp.*)
Milkbark (*Drypetes diversifolia*)
milkweed family (*Asclepias spp.*)
Milkweed, Common (*Asclepias syriaca*)
Milkweed, Mexican (*Asclepias currasavica*)
Milkweed, Sandhill (*Asclepias humistrata*)
Milkweed, Swamp (*Asclepias incarnata*)
Milkweed, White Swamp (*Asclepias perennis*)

Milkvetch, Canadian (*Astragalus canadensis*)
milkvetch (*Astragalus spp.*)
Mistletoe (*Phoradendron serotinum*)
mustard family (*Brassicaceae*)
Mustard, Garlic (*Alliaria officinalis*)
Mustard, Wild (*Brassica campestris*)
Myrtle, Wax (*Myrica cerifera*)
Nettle, Canadian Wood (*Laportea canadensis*)
Nettle, False (*Boehmeria cylindrica*)
Nettle, Stinging (*Urtica dioica*)
nettles (*Urtica spp.*)
nickerbean (*Caesalpinia spp.*)
Oak, Bear (*Quercus ilicifolia*)
Oak, Black (*Quercus velutina*)
Oak, Blackjack (*Quercus marilandica*)
Oak, Live (*Quercus virginiana*)
Oak, Myrtle (*Quercus myrtifolia*)
Oak, Northern Red (*Quercus rubra*)
Oak, Post (*Quercus stellata*)
Oak, Red (*Quercus rubra*)
Oak, Scrub (*Quercus ilicifolia*)
Oak, Southern Red (*Quercus falcata*)
Oak, Swamp White (*Quercus bicolor*)
Oak, Turkey (*Quercus laevis*)
Oak, Water (*Quercus nigra*)
Oak, White (*Quercus alba*)
oaks (*Quercus spp.*)
Oatgrass, Poverty (*Danthonia spicata*)
Orchardgrass (*Dactylis glomerata*)
panic grasses (*Panicum spp.*)
Paspalum, Rustyseed (*Paspalum langei*)
Passion Flower, Corky-stemmed (*Passiflora suberosa*)
Passion-Flower, Incense (*Passiflora "Incense"*)
Passion Flower, Many-flowered (*Passiflora multiflora*)
Passion Flower, Yellow (*Passiflora lutea*)
passion flowers (*Passiflora spp.*)

pawpaws (Asimina spp.)
Pea, Butterfly (Clitoria mariana)
Pea, Partridge (Cassia fasciculata)
Pea, Sensitive (Cassia nictitans)
Pea, Veiny (Lathyrus venosus)
Peanut, Hog (Amphicarpa bracteata)
Pellitory (Parietaria floridana)
Peppergrass, Field (Lepidium campestre)
Peppergrass, Virginia (Lepidium virginicum)
Pepperweed, Common (Lepidium densiflorium)
Petunia, Carolina Wild (Ruellia caroliniensis)
petunia, wild (Ruellia spp.)
Pine, Eastern White (Pinus strobus)
Pine, Loblolly (Pinus taeda)
Pine, Pitch (Pinus rigida)
Pine, Shortleaf (Pinus echinata)
Pine, Virginia (Pinus virginiana)
Pipevine, Elegant (Aristolochia elegans)
Pipevine, Wooly (Aristolochia tomentosa)
pipevines (Aristolochia spp.)
Plantain, Narrowleaf (Plantago lanceolata)
plantains (Plantago spp.)
Plum, American (Prunus americana)
Plum, Chickasaw (Prunus angustifolia)
Plum, Guiana (Drypetes lateriflora)
Plumbago, Wild (Plumbago scandens)
Plumegrass, Silver (Erianthus alopecuroidum)
poplars (Populus spp.)
pussy-toes (Antennaria spp.)
Radish, Wild (Raphanus raphanistrum)
Ragweed, Great (Ambrosia trifida)
Redbud (Cercis canadensis)
Redcedar, Eastern (Juniperus virginiana)
Reed, Common (Phragmites australis)

Rice, Annual Wild (Zizania aquatica)
rockcress (Arabis spp.)
rose family (Rosacea)
Saltwort (Batis maritima)
Sassafras (Sassafras albidum)
Sawgrass (Cladium jamaicense)
Sea Rocket (Cakile lanceolata)
sedge family (Cyperaceae)
Sedge, Shoreline (Carex hyalinolepis)
Sedge, Upright (Carex stricta)
sedges (Carex spp.)
Senna, Bahana (Cassia chapmanii)
Senna, Christmas (Cassia bicapsularis)
Senna, Coffee (Cassia occidentalis)
Senna, Maryland (Cassia marilandica)
Senna, Sicklepod (Cassia obtusifolia)
Senna, Wild (Cassia hebecarpa)
serviceberry (Amelanchier spp.)
Shepherd's Purse (Capsella bursa-pastoris)
Shorthusk, Bearded (Brachyelytrum erectum)
Sida (Sida acuta)
snake-herb (Dyschoriste spp.)
Snakeroot, Virginia (Aristolochia serpentaria)
Sneezeweed (Helenium autumnale)
Sorrel, Sheep (Rumex acetosella)
Spanish Dagger (Yucca aloifolia)
Spanish Needles (Bidens alba)
Sparkleberry (Vaccinium arboreum)
Spicebush (Lindera benzoin)
Sugarberry (Celtis laevigata)
Sumac, Fragrant (Rhus aromatica)
Sumac, Staghorn (Rhus typhina)
Sumac, Winged (Rhus copallina)
sumacs (Rhus spp.)
Summer Farewell (Dalea pinnata)
sunflowers (Helianthus spp.)
Sweet Everlasting (Gnaphalium obtusifolium)
Sweetleaf, Common (Symplocos

tinctoria)
Switchcane (Arundinaria gigantea var.tecta)
Switchgrass (Panicum virgatum)
Tamarind, Wild (Lysiloma latisiliquum)
Tea, Mexican (Chenopidium ambrosiodes)
Tea, New Jersey (Ceanothus americana)
Thistle, Yellow (Cirsium horridulum)
thistles (Cirsium spp.)
Threeawn, Arrowfeather (Aristida purpurascens var. virgata)
toadflax (Linaria spp.)
Toothwort, Cut-leaf (Dentaria laciniata)
toothworts (Dentaria spp.)
Torchwood (Amyris elemifera)
Tulip Tree (Liriodendron tulipifera)
Turtlehead (Chelone glabra)
twinflower (Dyschoriste spp.)
vegetables (broccoli, cabbage and others)
Velvetgrass, Common (Holcus lanatus)
Vetch, Carolina (Vicia caroliniana)
Vetch, Crown (Coronilla varia)
Vetch, Joint (Aeschyomene americana)
vetch, milk (Astragalus spp.)
vetches (Vinca spp.)
viburnum (Viburnum spp.)
Violet, Bird's-foot (Viola pedata)
violets (Viola spp.)
Walnut, Black (Juglans nigra)
walnuts (Juglans spp.)
Whitegrass (Leersia virginica)
Willow, Black (Salix nigra)
Willow, Carolina (Salix caroliniana)
Willow, Weeping (Salix babylonica)
willows (Salix spp.)
Wing-stem (Verbesina alternifolia)
Wisteria, American (Wisteria frutescens)
wisterias (Wisteria spp.)
Woodoats, Indian (Chasmanthium

latifolium)
Woolgrass (Scirpus cyperinus)
Yellowrocket, Garden (Barbarea vulgaris)
yuccas (Yucca spp.)

# CHECKLIST/INDEX

*Use the boxes to check the butterflies you've seen.*

## ABOUT THE AUTHOR

Jaret C. Daniels, Ph.D., is a professional nature photographer and entomologist at the University of Florida specializing in the ecology and conservation biology of Lepidoptera. He has authored numerous scientific papers, popular articles and books on butterflies, insects, wildlife conservation and butterfly gardening. He currently lives in Gainesville, Florida, with his wife Stephanie.